C++ Neural Networks and Fuzzy Logic

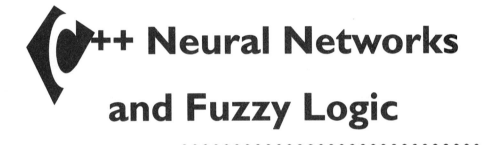

Valluru B. Rao and
Hayagriva V. Rao

MIS: PRESS

A Subsidiary of
Henry Holt and Co., Inc.

First Edition—1993

ISBN 1-55828-298-x

Printed in the United States of America.

10 9 8 7 6 5 4 3 2 1

MIS:Press books are available at special discounts for bulk purchases for sales promotions, premiums, fund-raising, or educational use. Special editions or book excerpts can also be created to specification.

For details contact: Special Sales Director
                      MIS:Press
                      a subsidiary of Henry Holt and Company, Inc.
                      115 West 18th Street
                      New York, New York 10011

Dedication

To the memory of
Vydehamma, Annapurnamma, Anandarao,
Madanagopalarao, Govindarao and Rajyalakshamma.

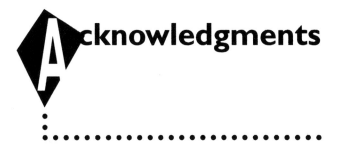Acknowledgments

We would like to thank Laura Lewin, Development Editor, Patty Wallenburg, Production Editor, and Kevin Latham of MIS:Press for bringing this project to fruition. We would also like to thank the Copy Editor, Suzanne Ingrao and the Technical Editor, Steve Gallant for all of the helpful comments and suggestions. Thanks also to Steve Berkowitz, Publisher, for believing in this project.

Last but not least, heartfelt thanks to Sarada, Rohini and Rekha for encouragement, support and putting up with us through another hectic project schedule.

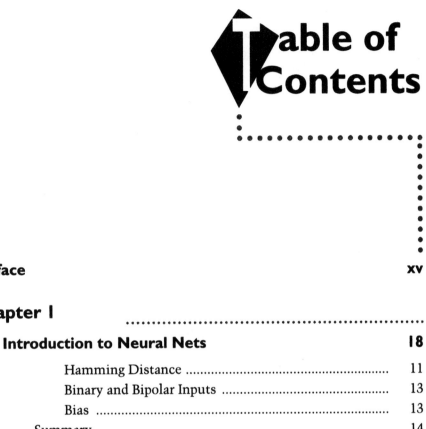

Table of Contents

Chapter 3

A Look at Fuzzy Logic   25

Chapter 4

Constructing a Neural Network   43

Chapter 5

Interactions: A Survey of Neural Network Models   **67**

Chapter 6

Chapter 7

Backpropagation 103

Chapter 8

BAM: Bidirectional Associative Memory 149

Chapter 11

The Kohonen Self-Organizing Map   223

Chapter 12

Application to Pattern Recognition   251

Chapter 13

Backpropagation II   267

Chapter 14

Application to Financial Forecasting   311

Preface

The number of models available in neural network literature is quite large. Very often the treatment is mathematical and complex. This book attempts to provide illustrative examples in C++ that the reader can use as a basis for further experimentation. One of the keys to learning about neural networks to appreciate their inner workings is to experiment. Neural networks, in the end, are fun to learn about and discover. Although the language for description used is C++, you will not find extensive class libraries in this book. With the exception of the backpropagation simulator, you will find fairly simple example programs for many different neural network architectures and paradigms. Since backpropagation is widely used and also easy to tame, a simulator is provided with the capacity to handle large input data sets. You use the simulator in one of the chapters in this book to solve a financial forecasting problem. You will find ample room to expand and experiment with the code presented in this book.

There are many different angles to neural networks and fuzzy logic. The fields are expanding rapidly with ever-new results and applications. This book presents many of the different neural network topologies, including the BAM, Hopfield memory, ART1, Kohonen's Self-organizing map, Kosko's Fuzzy Associative Memory, and, of course, Backpropagation. You should get a fairly broad picture of Neural Networks and Fuzzy Logic with this book. At the same time, you will have real code that shows you example usage of the models, to solidify your understanding. This is especially useful for the more complicated neural network architectures like the Adaptive Resonance Theory of Stephen Grossberg (ART).

The subjects are covered as follows:

◆ **Chapter 1** gives you an overview of Neural network terminology and nomenclature. You discover that neural nets are capable of solving complex problems with parallel computational architectures.

◆ **Chapter 2** introduces C++ and object orientation. You learn the benefits of object-oriented programming and basic concepts.

◆ **Chapter 3** introduces Fuzzy logic, a technology that is fairly synergistic with neural network problem solving. You learn about math with fuzzy sets as well as how you can build a simple fuzzifier in C++.

◆ **Chapter 4** introduces you to many aspects of neural networks in detail: the Hopfield network, the Perceptron network, and their C++ implementations.

◆ **Chapter 5** describes the features of several models, and threshold functions and develops concepts in neural networks.

◆ **Chapter 6** introduces the concepts of supervised and unsupervised learning and self-organization.

◆ **Chapter 7** goes through the construction of a backpropagation simulator. You will find this simulator useful in later chapters also . C++ classes and features are detailed in this chapter.

◆ **Chapter 8** covers the Bidirectional Associative Memories for associating pairs of patterns.

◆ **Chapter 9** introduces Fuzzy Associative Memories for associating pairs of fuzzy sets.

◆ **Chapter 10** covers the Adaptive Resonance Theory of Grossberg. You will have a chance to experiment with a program that illustrates the working of this theory.

◆ **Chapter 11** and **Chapter 12** discuss the Self-Organizing map of Teuvo Kohenen and its application to the problem of pattern recognition.

◆ **Chapter 13** continues the discussion of the backpropagation simulator, with enhancements made to the simulator.

- ◆ **Chapter 14** applies backpropagation to the problem of financial forecasting, discusses setting up a backpropagation network with 27 input variables and 200 test cases to run a simulation. The problem is approached via a systematic method for preprocessing data and careful analysis of training set and test set results to know when you've succeeded.

- ◆ **Chapter 15** deals with nonlinear optimization with a thorough discussion of the Traveling Salesperson Problem. You learn the formulation by Hopfield, and the approach of Kohonen.

Neural networks is now a subject of interest to professionals in many fields, and it is also a tool for many areas of problem solving. The applications are widespread in recent years, and the fruits of these applications are being reaped by many from diverse fields. This methodology has become an alternative to modeling of some physical and nonphysical systems with scientific or mathematical basis, and also to expert systems methodology. One of the reasons for it is that absence of full information is not as big a problem in neural networks as it is in the other methodologies mentioned earlier. The results are sometimes very astounding, and even phenomenal, with neural networks, and the effort is at times relatively modest to achieve such results. Image processing, vision, financial market analysis, optimization are among the many areas of application of neural networks. To think that the modeling of neural networks is one of modeling a system that attempts to mimic human learning is somewhat exciting. Neural networks can learn in in unsupervised learning mode. Just as human brains can be trained to master some situations, neural networks can be trained to recognize patterns and to do optimization and other tasks.

In the early days of interest in neural networks, the researchers were mainly biologists and psychologists. Serious research now is done by not only biologists and psychologists, but by professionals from electrical engineering, mathematics, and physics as well. The latter have either joined forces, or are doing independent research parallel with the former, who opened up a new and promising field for everyone.

In this book, we aim to introduce the subject of neural networks as directly and simply as possible for an easy understanding of the methodology. We also include the concept of fuzziness in neural networks. Most of the important neural network architectures are covered, and we earnestly hope that our efforts have succeeded in presenting this subject matter in a clear and useful fashion.

We welcome your comments and suggestions for this book, from errors and oversights, to suggestions for improvements to future printings at the following address:

H. Rao
P.O. Box 386
Plainsboro, N.J. 08536

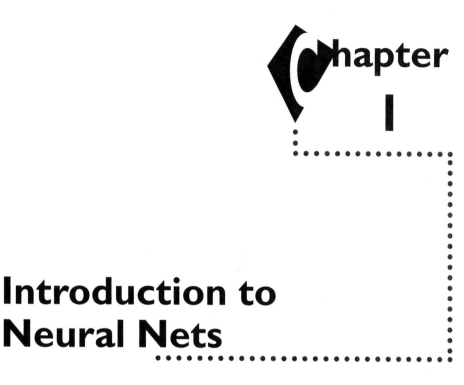

Chapter
1

Introduction to
Neural Nets

Neural Processing

How do you recognize a face in a crowd? How does an economist predict the direction of interest rates? Faced with problems like these, the human brain uses a web of interconnected processing elements called *neurons* to process information. Each neuron is autonomous and independent. It does it's work *asynchronously*, that is, without any synchronization to other events taking place. The problems described have two important characteristics that distinguish them from other problems: First, the solution to the problems are nonalgorithmic, that is, you can't devise a step-by-step *algorithm* or logical recipe to give you an answer; and second, the data provided to the problems is complex and may be noisy or incomplete. You could have forgotten your glasses when you're trying to recognize that face. The economist may have at his disposal thousands of pieces of data that may or may not be relevant to his forecast on the economy and on interest rates.

The vast processing power inherent in biological neural structures has inspired the study of the structure itself for hints on organizing man-made computing structures. *Artificial* neural networks, the subject of this book, covers the way to organize synthetic neurons to solve the same kind of difficult, nonalgorithmic problems in the same manner as we think the human brain may. This chapter will give you a sampling of the terms and nomenclature used to talk about neural networks. These terms will be covered in more depth in the chapters to follow.

Neural Network

A *neural network* is a group of processing elements where typically one subgroup makes independent computations and passes the results to a second subgroup. Each subgroup may in turn make its independent computations and pass on the results to yet another subgroup. Finally, a subgroup of one or more processing elements determines the output from the network. Each processing element makes its computation based upon a weighted sum of its inputs. A subgroup of processing elements is called a *layer* in the network. The first layer is the *input* layer and the last the *output* layer. The layers that are placed between the first and the last layers are the *hidden* layers. The processing elements are seen as units that are similar to the neurons in a human brain, and hence, they are referred to as *cells, neuromimes,* or *artificial neurons.* A *threshold function* is sometimes used to determine the output of a neuron in the output layer. The value outputted from the threshold function characterizes the neuron to have fired or not. Even though our subject matter deals with artificial neurons, we will simply refer to them as neurons, for brevity. Synapses between neurons are referred to as *connections,* which are represented by edges of a directed graph in which the nodes are the artificial neurons.

Figure 1.1 is a typical neural network. The circular nodes represent neurons. Here there are three layers, one input layer, a hidden layer, and an output layer. The directed graph mentioned shows the connections from layer to layer. You will see many variations on the number and types of layers throughout this book.

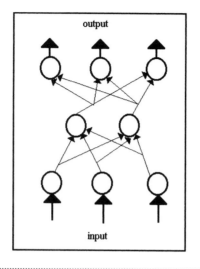

FIGURE 1.1 *A typical neural network.*

One application for a neural network is *pattern matching.* The patterns can consist of binary digits in the discrete cases, or real numbers representing analog signals in continuous cases. Pattern matching is a mode of pattern recognition and a form of establishing an association, called *autoassociation* or *heteroassociation*, depending on whether the associated pair consists of the same one or two different patterns. Pairing different patterns is building the type of association called heteroassociation. If you input a pattern A, and receive pattern A, this is termed autoassociation. What use does this provide? Remember the example given at the beginning of this chapter. Say you want to recall a face in a crowd and you have a hazy remembrance (input). What you want is the actual image. Autoassociation, then, is useful to recognize or retrieve patterns with possibly incomplete information as input. What about heteroassociation? Here you associate A with B. Given A, you get B and vice versa. You could store the face of a person and retrieve it with the person's name, for example. It's quite common in real circumstances to do the opposite, and sometimes not so well. You recall the face of a person, but can't place the name.

Once neural networks are constructed to do pattern matching and are analyzed, their capabilities in other directions and their limitations are well understood. That precise formulation and full knowledge of the system are not necessarily needed, that the neural networks can be trained, and that the networks can in some sense learn, are the realizations from the experience of initial development of neural networks. Processing in neural networks is done in parallel, rather than sequentially, as would be necessary in computing on digital computers. Computing on digital computers also requires precise information and algorithms. These observations show the greater potential of neural networks in problem solving, compared to expert systems, which also require precise formulations of firing rules based on full and precise information. It is no surprise that many different models of neural networks have been developed by researchers from varied fields, and the applications encompass a large spectrum, from image processing and hand writing analysis to optimization and control.

Output of a Neuron

Basically, the activation or raw output of a neuron in a neural network is a weighted sum of its inputs, but a threshold function is also used to determine the final value, or the output. When the output is 1, the neuron is said to *fire*, and when it is 0, the neuron is considered to have not fired. When a threshold function is used, different results of activations, all in the same interval of values, can cause the same final output value. This situation helps in the sense that if the neuron is going to fire for an activation value of, say 9, then even if you get the activation to be 10 and not 9, the neuron is going to fire. To put it in a simple and familiar setting, let us describe an example about a daytime game show on television, called the *Price is Right*.

Cash Register Game

One of the games a contestant in *Price is Right* is sometimes asked to play, is what they call the Cash Register game. A few products are described, their prices are unknown to the contestant, and the contestant has to declare how many units of each item he/she would like to (pretend to) buy. If the total purchase does not exceed the amount specified , the contestant wins some special prize. After the contestant announces how many items of a particular product he/she wants, the price of that product is revealed, and it is rung up on the cash register. The contestant must be careful, in this case, that the total does not exceed $7, to earn the associated prize. We can now cast the whole operation of this game, in terms of a neural network, called a *Perceptron*, as follows.

Consider each product on the shelf to be a neuron in the input layer, with its input being the unit price of that product. The cash register is the single neuron in the output layer. The only connections in the network are between each of the neurons (products displayed on the shelf) in the input layer and the output neuron (the cash register). This type of arrangement is usually referred to as a neuron, the cash register in this case, being an *instar*. The contestant actually wields these connections, because when the contestant says he/she wants, say 5, of a specific product, the contestant is thereby assigning a weight of 5 to the connection between that product and the cash register. The total bill for the purchases by the contestant is nothing but the weighted sum of the unit prices of the different products offered. For those items the contestant does not choose to purchase, the implicit weight assigned is 0. The application of the limit of $7 to the bill, is just the application of a threshold, except that the threshold value should not be exceeded for the outcome from this network to favor the contestant, winning him/her a good prize. Generally though, an output neuron is supposed to fire if its activation value exceeds the threshold value.

One observation can be made as a postscript to the above analogy. In this Cash Register game, the input neurons communicate with the output neuron *asynchronously*, meaning not all at once. Once one product is decided upon for its quantity by the contestant, it is already rung up on the cash register. In a neural network, the neurons in a single layer communicate to the neurons that they are connected to in a *synchronous* fashion.

Weights

The weights used on the connections between different layers have much significance in the working of the neural network and the characterization of a network. The following actions are possible in a neural network:

- ◆ Start with one set of weights and run the network.
- ◆ Start with one set of weights, run the network, and modify some or all the weights, and run the network again with the new set of weights.

◆ Repeat this process until some predetermined goal is met.

Training

The reason to alter the weights can be that the output(s) is not what is expected. Some rule then needs to be used to determine how to alter the weights. There should also be a criterion to specify when the process of successive modification of weights ceases. This process of changing the weights, or rather, refining the weights, is called *learning*. A network in which learning is employed is said to be subjected to *training*. Training is an external process or regimen. Learning is the desired result that takes place internal to the network.

Feedback

If you wish to train a network so it can learn some predetermined patterns, or some function values for given arguments, it would be important to have information fed back from the output neurons to neurons in some layer before that, to enable further processing and adjustment of weights on the connections. Such feedback can be to the input layer or a layer between the input layer and the output layer, sometimes labeled as the hidden layer. What is fed back is usually the error in the output, modified appropriately according to some useful paradigm. The process of feedback continues through the subsequent cycles of operation of the neural network and ceases when the training is completed.

Supervised/Unsupervised Learning

A network can be given supervised or unsupervised learning. The learning would be supervised if external criteria are used to be matched by the network output, and if not, unsupervised. This is one broad way to divide different neural network approaches. Unsupervised approaches are also termed *self-organizing*. There is more interaction between neurons, typically with feedback and intralayer connections between neurons playing a role in promoting self-organization. Supervised networks are quite straightforward. You apply the inputs to the network along with an expected response, much like the Pavlovian conditioned stimulus and response regimen. A stock market forecaster may present data to the neural network to the present day and hope to predict the future once training is complete.

Noise

A data set used to train a neural network may have inherent noise in it, or an image may have random speckles in it, for example. The response of the neural network to noise is an important factor in determining its suitability to a given application. You may wish to introduce noise intentionally in training to find out if the network is indeed learning in the presence of noise, and if the network is converging.

Memory

Once you train a network on a set of data, suppose you continue training the network with new data. Will the network forget the intended training on the original set, or will it remember? This is another angle that is approached by some researchers who are interested in preserving a network's *long-term memory* (LTM) as well as its *short-term memory* (STM).

Neural Network Construction

There are three aspects to the construction of a neural network. The first is *structure*. This relates to how many layers the network should contain, and what their functions are, such as for input, for output, etc. Structure also encompasses how interconnections are made between neurons in the network, and what their functions are. The second aspect is *encoding*. Encoding refers to the paradigm used for the determination of and changing of weights on the connections between neurons. Finally, *recall* is also an important aspect of a neural network. Recall refers to getting an expected output for a given input. If the same input as before is presented to the network, the same corresponding output as before should result. The type of recall can characterize the network as being autoassociative or heteroassociative. Autoassociation is the phenomenon of associating an input vector with itself as the output, whereas heteroassociation is that of recalling a related vector given an input vector. The three aspects to the construction of a neural network mentioned above essentially distinguish between different neural networks and are part of their design process. The architecture used defines the topology. Some relevant algorithms define the encoding process.

Cooperation and Competition

If the network consists of a single, input layer and an output layer consisting of a single neuron, then the set of weights for the connections between the input layer neurons and the output neuron are given in a weight vector. When the output layer has more than one neuron, the output is not just one value but a vector. In such a situation each neuron in one layer is interconnected with each neuron in the next layer, with weights assigned to these interconnections. Then the weights can all be given together in a weight matrix, which sometimes is referred to as the *correlation* matrix. When there are in-between layers such as a hidden layer or a so-called Kohonen layer or a Grossberg layer, the interconnections are made between each neuron in one layer and every neuron in the adjacent layer, and there will be a corresponding correlation matrix. Cooperation or competition or both can be imparted between network neurons in the same layer, through the right choice of weights in terms of their sign, but not in terms of their magnitude. Cooperation is the attempt between neurons in one neuron aiding the prospect of firing by another. Competition is the attempt between neurons to individually excel with higher output. Inhibition, a mechanism used in competition, is the attempt between neurons in one neuron decreasing the prospect of another neuron's firing. As already stated, the vehicle for these phenomena is the connection weight. For example, a positive weight is assigned for a connection between one node and a cooperating node in that layer, while a negative weight is assigned to inhibit a competitor.

To take this idea to the connections between neurons in consecutive layers, we would assign a positive weight to the connection between one node in one layer and its nearest neighbor node in the next layer, whereas the connections with distant nodes in the other layer will get negative weights. The negative weights would indicate competition in some cases and inhibition in others. To make at least some of the discussion and the concepts in it a bit clearer, we present an example.

Example—A Hopfield Network

The neural network we present is a Hopfield network, with a single layer. We place, in this layer, four neurons, each connected to the rest, as shown in Figure 1.2. Some of the connections have a positive weight, and the rest have a negative weight. The network will be presented with two input patterns, one at a time, and it is supposed to recall them. The inputs would be *binary* patterns having in each component a 0 or 1. If two patterns of equal length are given and are treated as vectors, their dot product is obtained by first multiplying corresponding components together and then adding these products. Two vectors are said to be orthogonal, if their dot product is 0. The mathematics involved in computations done for neural networks include matrix multiplication, transpose of a matrix, and transpose of a vector. Also see Appendix B. The inputs to be given should be *orthogonal* to one another.

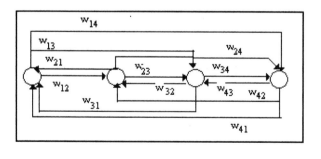

FIGURE 1.2 *Layout of a Hopfield Network.*

The two patterns we want the network to recall are A = (1, 0, 1, 0) and B = (0, 1, 0, 1), which you can verify to be orthogonal. Recall that two vectors A and B are orthogonal if their dot product is equal to zero. This is true in this case since $A_1B_1 + A_2 B_2 + A_3B_3 + A_4B_4 = (1 \times 0 + 0 \times 1 + 1 \times 0 + 0 \times 1) = 0$. The matrix W below gives the weights on the connections in the network.

$$W = \begin{matrix} 0 & -3 & 3 & -3 \\ -3 & 0 & -3 & 3 \\ 3 & -3 & 0 & -3 \\ -3 & 3 & -3 & 0 \end{matrix}$$

We need a threshold function also, and we define it as follows. The threshold value θ is 0.

$$f(t) = \begin{cases} 1 & \text{if } t >= \theta \\ 0 & \text{if } t < \theta \end{cases}$$

We have four neurons in the only layer in this network. We need to compute the activation of each neuron as the weighted sum of its inputs. The activation at the first node is the dot product of the input vector and the first column of the weight matrix. We get the activation at the other nodes similarly. The output of a neuron is then calculated by evaluating the threshold function at the activation of the neuron. So if we present the input vector A, the dot product works out to 3 and f(3) = 1. Similarly, we get the dot products of the second, third, and fourth nodes to be –6, 3, and –6, respectively. The corresponding outputs therefore are 0, 1, and 0. This means that the output of the network is the vector (1, 0, 1, 0), same as the input pattern. The network has recalled the pattern as presented. When B is presented, the dot product obtained at the first node is –6 and the output is 0. The outputs for the rest of the nodes taken together with the output of the first node gives (0, 1, 0, 1), which means that the network recalled B also. The weight matrix worked well, and we do not have to modify it.

What will the network give as output if we present a pattern different from both A and B? Let C = (0, 1, 0, 0) be presented to the network. The activations would be –3, 0, –3, 3, making the outputs 0, 1, 0, 1, which means that B is recalled. This is quite interesting. Suppose we did intend to input B and we made a slight error and ended up presenting C, instead. The network did alright and recalled B. But why not A? To answer this, let us ask is C closer to A or B? How do we compare? We use the distance formula for two four-dimensional points. If (a, b, c, d) and (e, f, g, h) are two four-dimensional points, the distance between them is the square root of:

$$(a-e)^2 + (b-f)^2 + (c-g)^2 + (d-h)^2$$

The distance between A and C is $\sqrt{3}$, whereas the distance between B and C is just 1. So since B is closer in this sense, B was recalled rather than A. You may verify that if we do the same exercise with D = (0, 0, 1, 0), we will see that the network recalls A, which is closer than B to D. On the other hand, if E = (1, 0, 0, 1), which is at an equal distance from A and B is used for input, the output will be F = (0, 1, 1, 0), which is neither A nor B. And again when (0, 1, 1, 0) becomes the input, E becomes the output.

In the discussion above, if you look upon C as a slightly perturbed B, or B affected by noise, the network coped with it and recalled B anyway. But E cannot be looked at in the same manner in reference to B, or to A for that matter. There are altogether 16 4-bit patterns, of which (1, 1, 0, 1), (0, 0, 0, 1), and (0, 1, 1, 1), are as close to B as C is. They all cause the network to recall B.

Hamming Distance

Talking about closeness of a bit pattern to another bit pattern, the Euclidean distance need not be considered. Instead, the *Hamming distance* can be used, which is much easier to determine, since it is the number of bit positions in which the two patterns being compared differ. Patterns being strings, Hamming distance is more appropriate than the Euclidean distance.

NOTE

The weight matrix W we gave in this example is not the only weight matrix that would enable the network to recall the patterns A and B correctly. You can see that if we replace each of 3 and –3 in the matrix by say, 2 and –2, respectively, the resulting matrix would also facilitate the same performance from the network.

Suppose we are interested in having the patterns E and F also recalled correctly, in addition to the patterns A and B. We need to train the network, and we need to come up with a learning algorithm. Suppose we did that and came up after a few refinements with the matrix W_1:

$$W_1 = \begin{array}{cccc} 0 & -5 & 4 & 4 \\ -5 & 0 & 4 & 4 \\ 4 & 4 & 0 & -5 \\ 4 & 4 & -5 & 0 \end{array}$$

Using this modification of the weight matrix, we are successful in having the network recall all four patterns A, B, E, and F.

The other patterns such as C and D, which are not in the training set of patterns to this network, can be presented to the network, and the results can be examined. C = (0, 1, 0, 0) will cause the output to be (0, 1, 1, 1), which is not only C, but it is not any of the training patterns either. In the case of D = (0, 0, 1, 0), the output pattern is (1, 1, 1, 0) deserving a similar comment as above.

Neither C nor D has a unique closest pattern among A, B, E, and F as you can see from Table 1.1.

TABLE 1.1 *Hamming distances.*

Pattern	Hamming distance from			
	A 1 0 1 0	**B** 0 1 0 1	**E** 1 0 0 1	**F** 0 1 1 0
C(0 1 0 0)	3	1	3	1
D(0 0 1 0)	1	3	3	1

Let us close this example by presenting the outputs for some other input patterns in Table 1.2.

TABLE 1.2 *Outputs and some input patterns.*

Input Pattern	Hamming distance from				Output Pattern
	A 1 0 1 0	**B** 0 1 0 1	**E** 1 0 0 1	**F** 0 1 1 0	
G(0 1 1 1)	3	1	3	1	H(1 1 0 0)
H(1 1 0 0)	2	2	2	2	I(0 0 1 1)
					Continued

Input Pattern	Hamming distance from				Output Pattern
	A 1 0 1 0	B 0 1 0 1	E 1 0 0 1	F 0 1 1 0	
I(0 0 1 1)	2	2	2	2	H(1 1 0 0)
J(1 1 1 0)	1	3	3	1	I(0 0 1 1)

It is interesting to note that the patterns H(1, 1, 0, 0) and I(0, 0, 1, 1) which have a Hamming distance of 4, the maximum possible value, between them are the outputs for each other. In the set of patterns we encountered here, A and B are orthogonal, E and F are orthogonal, and H and I are orthogonal. The Hamming distance between these pairs of orthogonal vectors is 4, the maximum possible value. Of these three pairs, the first pair, namely A and B, is used to determine the weight matrix and the network makes autoassociation for each of these patterns. The other two pairs of orthogonal patterns are heteroassociated by the network. Finally, note also that input C gave output G, while input D gave output J.

Binary and Bipolar Inputs

Two types of inputs that are used in neural networks are binary and bipolar inputs. We have already seen examples of binary input. Bipolar inputs have one of two values, 1 and -1. There is clearly a one-to-one mapping or correspondence between them, namely having -1 of bipolar correspond to a 0 of binary. In determining the weight matrix in some situations where binary strings are the inputs, this mapping is used, and when the output appears in bipolar values, the inverse transformation is applied to get the corresponding binary string. A simple example would be that the binary string 1 0 0 1 is mapped onto the bipolar string 1 –1 –1 1; while using the inverse transformation on the bipolar string –1 1 –1 –1, we get the binary string 0 1 0 0.

Bias

The use of threshold value can take two forms. One is what we showed in the example. The activation is compared to the threshold value and the neuron fires if the threshold value is attained or exceeded. The other way is to add a value to the activation itself, in which case it is called the bias, and then determining the output of the neuron. We will encounter bias and gain later.

Summary

In this chapter we introduced a neural network as a collection of processing elements distributed over a finite number of layers and interconnected with positive or negative weights, depending on whether cooperation or competition (or inhibition) is intended. The activation of a neuron is basically a weighted sum of its inputs. A threshold function determines the output of the network. There may be layers of neurons in between the input layer and the output layer, and some such middle layers are referred to as hidden layers, others by names such as Grossberg or Kohonen layers, named after the researchers Stephen Grossberg and Teuvo Kohonen, who proposed them and their function. Modification of the weights is the process of training the network, and a network subject to this process is said to be learning during that phase of the operation of the network. In some network operations, a feedback operation is used in which the current output is treated as modified input to the same network and is fed back to the network through the hidden layer. This context makes the purpose of the presence of the hidden layer clearer.

Neural networks can be used for problems with nonalgorithmic solutions and for problems with incomplete or noisy data.

Chapter 2

C++ and Object Orientation

Introduction to C++

C++ is an *object-oriented* programming language built on the base of the C language. This chapter gives you a brief introduction to C++, touching on many important aspects of C++, so you would be able to follow our presentations of the C++ implementations of neural network models and write your own C++ programs.

Object-oriented programming is different from conventional programming in that data items and functions that manipulate those data items are bundled together and deployed wherever they are needed. Such bundles are called *objects*, and their data structures are called *classes*. Some of the headaches for programmers in conventional programming are in the need to prevent duplication of names of data items and functions, in making sure

15

that argument types are correct in function calls and so on. These types of problems can be avoided with C++ object-oriented programming. Therefore, objects created by one programmer for one particular application are usable by other programmers if the same objects are relevant for their applications. This reuse of code relating to the creation and implementation of objects can be done by the other programmers, without worrying about how variables and functions are named in the rest of the code application.

Object-oriented programming provides the ease of extension of code and reuse of code as well. For large projects of software development, object-oriented programming is a tool enhancing the efficiency and reliability of code. There is of course a greater effort in the planning stages of the code generation with object-oriented programming compared with traditional programming. This is a small price to pay to receive the greater modularity and other benefits from object-oriented programming.

Encapsulation

In C++ you have the facility to encapsulate data and the operations that manipulate that data, in an appropriate object. This enables the use of these objects in programs other than for which they were originally created. The encapsulation of data and the intended operations on them also prevents the data from being subjected to operations not meant for them. The operations are usually given in the form of functions using the data items as arguments. Such functions are also called *methods* in some object-oriented programming languages. The data items and the functions that manipulate them are combined into a data structure called *class*. A class is an abstract data type.

Hiding Data

Encapsulation also hides the data from other classes and functions in other classes. In C++ the access to an object, and its encapsulated data and functions is treated very carefully, by the use of keywords *private*, *protected*, and *public*. One has the opportunity to make access specifications for data objects and functions as being private, or protected, or public while defining a class. Only when the declaration is made as public do other functions and objects have access to the object and its components without question. On the other hand, if the declaration happens to be as private, there is no possibility of such access. When the declaration given is as protected, then the access to data and functions in a class by others is not as free as when it is public, nor as restricted as when it is private. You can declare one class as derived from another class. So-called derived classes and the declaring class do get the access to the components of the object that are declared protected. One class that is not a derived class of a second class can get access to data items and functions of the second class if it is declared as a *friend class* in the second. The three types of declarations of access specification can be different for different components of an object. For example, some of the data items could be declared public, some private, and the others protected. The same situation can occur with the functions in an object. When no explicit declaration is made, the default specification is as private. This aspect of encapsulation together with that of control of access makes C++ an object-oriented programming language.

Constructors and Destructors as Special Functions of C++

Constructors and destructors are special functions in C++. You cannot have a class defined in a C++ program without declaring and defining at least one constructor for it. You may omit declaring and then defining a destructor only because the compiler you use will create a default destructor. More than one constructor, but only one destructor, can be declared for a class.

Constructors are for the creation of an object of a class and for initializing it. C++ requires that every function has a return type. The only exceptions are constructors and destructors. A constructor is given the same name as the class for which it is a constructor. It may take arguments or it may not need them. Different constructors for the same class differ in the number and types of arguments they take. It is a good idea to provide for each class at least a default constructor that does not take any arguments and does not do anything except create an object of that class type. A constructor is called at the time an object of its class is needed to be created.

A destructor also is given the same name as the class for which it is a destructor, but with the *tilde* (~) preceding the name. Typically, what is done in a destructor is to have statements that ask the system to delete the various data structures created for the class. This helps to free-up allocated memory for those data structures. A destructor is called when the object created earlier is no longer needed in the program.

Dynamic Memory Allocation

C++ has keywords *new* and *delete*, which are used as a pair in that order, though separated by other statements of the program. They are for making dynamic allocation of memory at the time of creation of a class object and for freeing-up such allocated memory when it is no longer needed. You create space on the heap with the use of new. This obviates the need in C++ for *malloc,* which is the function for dynamic memory allocation used in C.

Overloading

Encapsulation of data and functions would also allow you to use the same function name in two different objects. The use of a name for a function more than once does not have to be only in different object declarations. Within the same object one can use the same name for functions with different functionalities, if they can be distinguished in terms of either their return type, or in terms of their argument types and number. This feature is called *overloading*. For example, if two different types of variables are data items in an object, a commonly named function can be the addition, one for each of the two types of variables—thus taking advantage of overloading. Then the function addition is said to be overloaded. But remember that the function **main** is just about the only function that cannot be overloaded.

Polymorphic Functions

A *polymorphic* function is a function whose name is used in different ways in a program. It can be also declared *virtual*, if the intention is *late binding*.

This causes it to be bound at run time. Late binding is also referred to as *dynamic* binding. An advantage in declaring a function in an object as virtual is that, if the program that uses this object calls that function only conditionally, there is no need to bind the function early, during the compilation of the program. It will be bound only if the condition is met and the call of the function becomes a fact. For example, you could have a polymorphic function called *draw()* that is associated with different graphical objects. The details or *methods* of the functions are different, but the name draw() is common. If you now have a collection of these objects and pick up an arbitrary object without knowing exactly what it is (via a pointer, for example), you can still invoke the draw function for the object and be assured that the right draw function will be bound to the object and called.

Overloading Operators

You can overload operators in addition to overloading functions. As a matter of fact, the system defined left shift operator << is also overloaded in C++ when used with **cout**, the C++ variation of the C language **printf** function. There is a similar situation with the right shift operator >> in C++ when used with **cin**, the C++ variation of the C language **scanf** function. You can take any operator and overload it. But you want to be cautious and not overdo it, and also you do not create confusion when you overload an operator. The guiding principle in this regard is the creation of a code-saving and time-saving facility while maintaining simplicity and clarity. Operator overloading is especially useful for doing normal arithmetic on nonstandard data types. You could overload the multiplication symbol to work with complex numbers, for example.

Inheritance

The primary distinction for C++ from C is that C++ has classes. Objects are defined as classes. Classes themselves can be data items in other classes, in which case one class would be an element of another class. Of course then one class is a member, which brings with it its own data and functions, in the second class.

A relationship between classes can be established not only by making one class a member of another, but also by the process of deriving one class from another. One class can be derived from another class, which becomes its base class. Then a hierarchy of classes is established, and a sort of parent–child relationship between classes is established. The derived class inherits, from the base class, some of the data members and functions. Naturally, if a class A is derived from a class B, and if B itself is derived from a class C, then A inherits from both B and C. A class can be derived from more than one class. This is how multiple inheritance occurs.

Derived Classes

When one class has some members declared in it as protected, then such members would be hidden from other classes, but not from the derived classes. In other words, deriving one class from another is a way of accessing the protected members of the parent class by the derived class. We then say that the derived class is inheriting from the parent class those members in the parent class that are declared as protected or public.

In declaring a derived class from another class, access or visibility specification can be made, meaning that such derivation can be public or the default case, private. Table 3.1 shows the consequences of such specification when deriving one class from another.

TABLE 3.1   *Visibility of base class members in derived class.*

Derivation Specification	Base class Specification	Derived class Access
private	private	none
(default)	protected	full access, private in derived class
	public	full access, public in derived class
public	private	none
	protected	full access, protected in derived class

Continued

Derivation Specification	Base class Specification	Derived class Access
	public	full access, public in derived class

Reuse of Code

C++ is also attractive for the extendability of the programs written in it and for the reuse opportunity, thanks to the features in C++ such as inheritance and polymorphism mentioned earlier. A new programming project can not only reuse classes that are created for some other program, if they are appropriate, but can extend another program with additional classes and functions as deemed necessary.

C++ Compilers

All of the programs in this book have been compiled and tested with Turbo C++, Borland C++, Microsoft C/C++ and Microsoft Visual C++. These are a few of the popular commercial C++ compilers available. You should be able to use most other commercial C++ compilers also.

Writing C++ Programs

Before one starts writing a C++ program for a particular problem, one has to have a clear picture of the various parameters and variables that would be part of the problem definition and/or its solution. In addition, it should be clear as to what manipulations need to be performed during the solution process. Then one carefully determines what classes are needed and what relationships they have to each other in a hierarchy of classes. It would be far more clear to the programmer at this point in the program plan what the data and function access specifications should be and so on. The typical compilation error messages a programmer to C++ may encounter are stating that a particular function or data is not a member of a particular class, or that it is not accessible to a class, or that a constructor was not available for a particular class. When function arguments at declaration and at the place the function is called do not match, either for number or for types or both, the compiler thinks of them as two different functions. The compiler does not find the definition and/or declaration of one of the two and has reason to complain. This type of error in one line of code may cause the compiler to alert you that several other errors are also present, perhaps some in terms of improper punctuation. In that case remedying the fundamental error that was pointed out would straighten many of the other argued matters.

The following list contains a few additional particulars you need to keep in mind when writing C++ programs.

◆ A member x of an object A is referred to with **A.x** just as done with structure elements in C.

◆ If you declare a class B, then the constructor function is also named B. B has no return type. If this constructor takes, say, one argument of type integer, you define the constructor using the syntax: B::B(int){whatever the function does};

◆ If you declare a member function C of class B, where return type of C is, say, float and C takes two arguments, one of type float, and the other int, then you define C with the syntax: float B::C(float,int){whatever the function does};

◆ If you declare a member function D of class B, where D does not return any value and takes no arguments, you define D using the syntax: void B::D( ){whatever the function does};

◆ If G is a class derived from, say, class B previously mentioned, you declare G using the syntax: class G:B. The constructor for G is defined using the syntax: G::G(arguments of G):B(int){whatever the function does}. If, on the other hand, G is derived from, say, classes B as well as T, then you declare G using the syntax: class G:B,T.

◆ If, one class is declared as derived from more than one other class, that is, if there are more than one base class for it, the derivations specification can be different or the same. Thus the class may be derived from one class publicly and at the same time from another class privately.

◆ If you have declared a global variable y external to a class B, and if you also have a data member y in the class B, you can use the external y with the reference symbol ::. Thus ::y refers to the global variable while y, within a member function of B, or B.y refers to the data member of B. This way polymorphic functions can also be distinguished from each other.

This is by no means a comprehensive list of features, but more a review of important constructs in C++. You will see examples of C++ usage in the chapters ahead.

Summary

A few highlights of the C++ language are presented.

◆ C++ is an object-oriented language with full compatibility with the C language.

◆ You create classes in C++ that encapsulate data and functions that operate on the data.

◆ You can create hierarchies of classes with the facility of inheritance. Polymorphism is a feature that allows you to apply a function to a task according to the object the function is operating on.

◆ Another feature in C++ is overloading of operators, which allows you to create new functionality for existing operators in a different context.

◆ Overall, C++ is a powerful language fitting the object-oriented paradigm that enables software reuse and enhanced reliability.

A Look at Fuzzy Logic

Crisp or Fuzzy Logic?

Logic deals with propositions that may or may not be true. A *proposition* can be true on one occasion, and false on another. "Apple is a red fruit" is such a proposition. If you are holding a Granny Smith apple that is green, the proposition that apple is a red fruit is false. On the other hand, if your apple is of a red delicious variety, it is a red fruit and the proposition in reference is true. If a proposition is true, it has a truth value of 1, and if it is false, its truth value is 0. These are the only possible truth values. Propositions can be combined to generate other propositions, by means of logical operations. The truth value of a proposition thus obtained can be related to the truth values of propositions that were combined to generate it.

When you say it will rain today, or that you can have a nice outdoor picnic today, you are making statements with certainty. Of course your

25

statements in this case can be either true, or false. The truth values of your statements can be only 1, or 0. Your statements then can be said to be *crisp*. On the other hand, there are statements you cannot make with such certainty. You may be saying that you think it will rain today. If pressed further you may be able to say with a degree of certainty in your statement that it will rain today. Your level of certainty, however, is about 0.8, rather than 1. In this case you are using *fuzzy* logic. Fuzzy logic deals with propositions that can be true to a certain degree—somewhere from 0 to 1. Therefore, a proposition's truth value indicates the degree of certainty about which the proposition is true. The degree of certainity sounds like a probability (perhaps subjective probability), but it is not quite the same. Probabilities for mutually exclusive events cannot add up to more than 1, but their fuzzy values could. For example, **hot** and **cold** can have 0.6 and 0.5 as their fuzzy values, but not as probabilities. Propositions can be combined to generate other propositions, using logical operations.

Fuzzy Sets

Fuzzy logic is best understood in the context of *set membership*. Suppose you are building a set of rainy days. Would you put today into the set? When you deal only with crisp statements, statements that are either true or false, your inclusion of today in the set of rainy days is based on certainty only. When dealing with Fuzzy Logic, you would include today in the set of rainy days via an *ordered pair*, such as (today, 0.8). The first member in such an ordered pair is a *candidate for inclusion* in the set, and the second member is a value between 0 and 1, inclusive, called the *degree of membership* in the set. The inclusion of the degree of membership in the set makes it convenient for developers to come up with a set theory based on fuzzy logic, just as regular set theory is developed. Fuzzy sets are sets in which members are presented as ordered pairs that include information on degree of membership. A traditional set of say, k elements, is a special case of a fuzzy set, where each of those k elements has 1 for the degree of membership, and every other element in the universal set has a degree of membership zero, for which reason you don't bother to list it.

Fuzzy Set Operations

The usual operations you can perform on ordinary sets are *union*, in which you take all of the elements that are in one set or the other; and the *intersection*, in which you take the elements that are in one and the other of two sets. In the case of fuzzy sets, your job while taking a union is to find the degree of membership that an element should have in the new fuzzy set, which is the union of two fuzzy sets.

If a, b, c, and d are such that their degrees of membership in the fuzzy set A are 0.9, 0.4, 0.5, and 0, respectively, then the fuzzy set A is given by the *fit vector*, (0.9, 0.4, 0.5, 0). The components of this fit vector are called *fit values* of a, b, c, and d.

Union of Fuzzy Sets

Consider a union of two traditional sets, and an element that belongs to only one of those two sets. You saw above that if you treat these sets as fuzzy sets, this element has a degree of membership of 1 in one case and 0 in the other since it belongs to one set and not to the other. Yet you are going to put this element in the union. The criterion you use in this action has to do with degrees of membership. You need to look at the two degrees of membership, namely 0 and 1, and pick the higher value of the two, namely 1. In other words, what you want for the degree of membership of an element when listed in the union of two fuzzy sets, is the *maximum* value of its degrees of membership within the two fuzzy sets that are forming a union.

If a, b, c, and d have the respective degrees of membership in fuzzy sets A, B as A = (0.9, 0.4, 0.5, 0) and B = (0.7, 0.6, 0.3, 0.8), then A U B = (0.9, 0.6, 0.5, 0.8).

Intersection and Complement of Two Fuzzy Sets

Analogously, the degree of membership of an element in the intersection of two fuzzy sets is the *minimum*, or the smaller value of its degree of membership individually in the two sets forming the intersection. For example, if today has 0.8 for degree of membership in the set of rainy days, and 0.5 for degree of membership in the set of days of work completion, then today belongs to the set of rainy days on which work is completed to a degree of 0.5, the smaller of 0.5 and 0.8.

Recall the fuzzy sets A and B in the previous example. A = (0.9, 0.4, 0.5, 0) and B = (0.7, 0.6, 0.3, 0.8). A∩B which is the intersection of the fuzzy sets A and B, is obtained by taking, in each component, the smaller of the values found in that component in A and in B. Thus A∩B = (0.7, 0.4, 0.3, 0).

The idea of a universal set is implicit in dealing with traditional sets. For example, if you talk of the set of married persons, the universal set is the set of all persons. Every other set you consider in that context is a subset of the universal set. We bring up this matter of universal set because when you make the complement of a traditional set A, you need to put in every element in the universal set that is not in A. The complement of a fuzzy set, however, is obtained as follows. In the case of fuzzy sets, if the degree of membership is 0.8 for a member, then that member is not in that set to a degree of 1.0 – 0.8 = 0.2. So you can set the degree of membership in the complement fuzzy set to the complement with respect to 1. If we return to the scenario of having a degree of 0.8 in the set of rainy days, then today has to have 0.2 membership degree in the set of nonrainy or clear days.

Continuing with our example of fuzzy sets A and B, and denoting the complement of A by A', we have A' = (0.1, 0.6, 0.5, 1) and B' = (0.3, 0.4, 0.7, 0.2). Note that A' U B' = (0.3, 0.6, 0.7, 1), which is also the complement of A ∩ B. You can similarly verify that the complement of A U B is the same as A' ∩ B'. Furthermore, A U A' = (0.9, 0.6, 0.5, 1) and A ∩ A' = (0.1, 0.4, 0.5, 0), which is not a vector of zeroes only, as would be the case in conventional sets. In fact A and A' will be equal in the sense that their fit vectors are the same, if each component in the fit vector is equal to 0.5.

Applications of Fuzzy Logic

Applications of fuzzy sets and fuzzy logic are found in many fields, such as artificial intelligence, engineering, computer science, operations research, robotics, and pattern recognition. These fields are also ripe for applications for neural networks. So it seems very natural that fuzziness should be introduced in neural networks themselves. Any area where humans need to indulge in making decisions, fuzzy sets can find a place, since information on which decisions are to be based may not always be complete, and the reliability of the supposed values of the underlying parameters is not always certain.

Examples of Fuzzy Logic

Let us say five tasks have to be performed in a given period of time, and each task requires one person dedicated to it. Suppose there are six people capable of doing these tasks. As you have more than enough people, there is no problem in scheduling this work and getting it done. Of course who gets assigned to which task depends on some criterion, such as total time for completion, on which some optimization can be done. But suppose these six people are not necessarily available during the particular period of time in question. Suddenly, the equation is seen in less than crisp terms. The availability of the people is fuzzy-valued. Here is an example of an assignment problem where fuzzy sets can be used.

Commercial Applications

Many commercial uses of fuzzy logic exist today. A few examples are listed below:

- ◆ A subway in Sendai, Japan uses a fuzzy controller to control a subway car. This controller has outperformed human and conventional controllers in giving a smooth ride to passengers in all terrain and external conditions.

- ◆ Cameras and camcorders use fuzzy logic to adjust autofocus mechanisms and also to cancel the jitter caused by a shaking hand.

- ◆ Some automobiles use fuzzy logic for different control applications. Nissan has patents on fuzzy logic braking systems, transmission controls and fuel injectors. GM uses a fuzzy transmission system in its Saturn vehicles.

Fuzziness in Neural Networks

There are a number of ways fuzzy logic can be used with neural networks. Perhaps the simplest way is to use a *fuzzifier* function to pre-process or post-process data for a neural network. This is shown in Figure 3.1, where a neural network has a pre-processing fuzzifier that converts data into fuzzy data for application to a neural network.

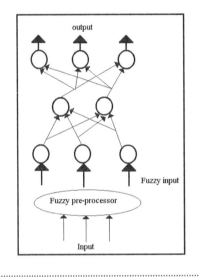

FIGURE 3.1   *A neural network with fuzzy preprocessor.*

Let us build a simple fuzzifier based on an application to predict the direction of the stock market. Suppose that you wish to fuzzify one set of data used in the network, the Federal Reserve's fiscal policy, in one of four fuzzy categories: very accomodative, accomodative, tight, or very tight. Let us suppose that the raw data that we need to fuzzify is the discount rate, and the interest rate that the Federal Reserve controls to set the fiscal policy. Now, a low discount rate usually indicates a loose fiscal policy, but this depends not only on the observer, but also on the political climate. There is a probability, for a given discount rate, you will find two people who offer different categories for the Fed fiscal policy. Hence it is appropriate to fuzzify the data, so that the data we present to the neural network is much as what an observer would see.

Figure 3.2 shows the fuzzy categories for different interest rates. Note that the category tight, has the largest range. At any given interest rate level, you could have one possible category or several. If only one interest rate is present on the graph, this indicates that membership in that fuzzy set is 1.0. If you have three possible fuzzy sets, there is a requirement that membership probability add up to 1.0. For an interest rate of 8%, you have some chance of finding this in the tight category or in the accomodative category. To find out the percentage probability from the graph, take the height of each curve at a given interest rate and normalize this to a one-unit length. At 8%, the tight category is about 0.8 unit in height, and acco-modative is about 0.3 unit in height. The total is about 1.1 units, and the probability of the value being tight is then .8/1.1 = .73, while the probabili-ty of the value being accommodative is .27.

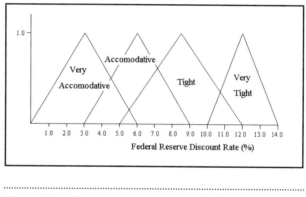

Figure 3.2 *Fuzzy categories for Federal Reserve Policy based on the Fed discount rate.*

Example Code

Let's develop C++ code to create a simple fuzzifier. A class is defined in Listing 3.1 called category. This class encapsulates the data that we need to define, those categories in Figure 3.2. There are three private data members called *lowval*, *midval* and *highval*. These represent the values on the graph that define the category triangle. In the tight category, the lowval is 5.0, the midval is 8.5 and the highval is 12.0. The category class allows you to instantiate a category object and assign parameters to it to define it. Also, there is a string called *name* that identifies the category, e.g. "tight".

Various member functions are used to interface to the private data members. There is **setval()**, for instance, that lets you set the value of the three parameters, while **gethighval()** returns the value of the parameter highval. The function **getshare()** returns the relative value of membership in a category given an input. In the example discussed previously, with the number 8.0 as the Fed discount rate, and the category tight defined according to the graph in Figure 3.2, **getshare()** would return 0.8. Note that this is not yet normalized. Following this example, the **getshare()** value from the accomodative category would also be used to determine the membership weights. These weights define a probability in a given category. A random number generator is used to define a value that is used to select a fuzzy category based on the probabilities defined.

Listing 3.1   fuzzfier.h

```
// fuzzfier.h    V. Rao, H. Rao
// program to fuzzify data


class category
{
private:
    char name[30];
    float  lowval,highval,midval;

public:
    category(){};
    void setname(char *);
    char * getname();
    void setval(float&,float&,float&);
    float getlowval();
    float getmidval();
    float gethighval();

    float getshare(const float&);

    ~category(){};

};

int randnum(int);
```

Let's look at the implementation file in Listing 3.2:

Listing 3.2   fuzzfier.cpp

```
// fuzzfier.cpp     V. Rao, H. Rao
// program to fuzzify data
```

```
#include <iostream.h>
#include <stdlib.h>
#include <time.h>
#include <string.h>
#include <fuzzfier.h>

void category::setname(char *n)
{
strcpy(name,n);
}

char * category::getname()
{
return name;
}

void category::setval(float &h, float &m, float &l)
{
highval=h;
midval=m;
lowval=l;
}

float category::getlowval()
{
return lowval;
}

float category::getmidval()
{
return midval;
}

float category::gethighval()
{
return highval;
}

float category::getshare(const float & input)
{
// this member function returns the relative membership
// of an input in a category, with a maximum of 1.0

float output;
float midlow, highmid;

midlow=midval-lowval;
highmid=highval-midval;

// if outside the range, then output=0
if ((input <= lowval) || (input >= highval))
```

```
        output=0;
    else
        {

        if (input > midval)

            output=(highval-input)/highmid;


        else

            if (input==midval)

                output=1.0;

            else

                output=(input-lowval)/midlow;

        }
    return output;
    }

int randomnum(int maxval)
{
// random number generator
// will return an integer up to maxval

srand ((unsigned)time(NULL));
return rand() % maxval;
}

void main()
{
// a fuzzifier program that takes category information:
// lowval, midval and highval and category name
// and fuzzifies an input based on
// the total number of categories and the membership
// in each category

int i=0,j=0,numcat=0,randnum;
float l,m,h, inval=1.0;

char input[30]="                ";
category * ptr[10];
float relprob[10];
float total=0, runtotal=0;

//input the category information; terminate with 'done';
```

```
while (1)

    {

    cout << "\nPlease type in a category name, e.g. Cool\n";
    cout << "Enter one word without spaces\n";
    cout << "When you are done, type 'done' :\n\n";

    ptr[i]= new category;
    cin >> input;


    if ((input[0]=='d' && input[1]=='o' &&
        input[2]=='n' && input[3]=='e')) break;

    ptr[i]->setname(input);

    cout << "\nType in the lowval, midval and highval\n";
    cout << "for each category, separated by spaces\n";
    cout << " e.g. 1.0 3.0 5.0 :\n\n";

    cin >> l >> m >> h;
    ptr[i]->setval(h,m,l);

    i++;

    }

numcat=i; // number of categories

// Categories set up: Now input the data to fuzzify
cout <<"\n\n";
cout << "======================================\n";
cout << "==Fuzzifier is ready for data==\n";
cout << "======================================\n";

while (1)

    {
    cout << "\ninput a data value, type 0 to terminate\n";

    cin >> inval;

    if (inval == 0) break;

    // calculate relative probabilities of
    // input being in each category

    total=0;

    for (j=0;j<numcat;j++)
```

```
        {
        relprob[j]=100*ptr[j]->getshare(inval);
        total+=relprob[j];
        }

    if (total==0)
        {
        cout << "data out of range\n";
        exit(1);
        }

    randnum=randomnum((int)total);

    j=0;
    runtotal=relprob[0];

    while ((runtotal<randnum)&&(j<numcat))
        {
        j++;
        runtotal += relprob[j];
        }

    cout << "\nOutput fuzzy category is ==> " <<
        ptr[j]->getname()<<"<== \n";

        cout <<"category\t"<<"membership\n";
        cout <<"---------------\n";

    for (j=0;j<numcat;j++)
        {
        cout << ptr[j]->getname()<<"\t\t"<<
            (relprob[j]/total) <<"\n";
        }

    }

cout << "\n\nAll done. Have a fuzzy day !\n";

}
```

This program first sets up all of the categories that you define. These could be for the example we choose or any example you can think of. After the categories are defined, you can start entering data to be fuzzified. As you keep entering data you see the probability aspect come into play. If you enter the same value twice, you may end up with different categories! You will see sample output shortly, but first a technical note on how the weighted probabilities are set up. The best way to explain it is with an example. Suppose that you have defined three categories, A, B and C.

Suppose that category A has a relative membership of 0.8, category B of 0.4 and category C of 0.2. In the program, these numbers are first multiplied by 100, so you end up with A=80, B=40 and C=20. Now these are stored in a vector with an index j initialized to point to the first category. Let's say that these three numbers represent three adjacent number bins that are joined together. Now pick a random number to index into the bin that has its maximum value of (80+40+20). If the number is 100, then it is greater than 80, and less than (80+40), you end up in the second bin which represents B. Does this scheme give you weighted probabilities? Yes it does, since the size of the bin (given a uniform distribution of random indexes into it) determines the probability of falling into the bin. Therefore, the probability of falling into bin A is 80/(80+40+20).

Sample output from the program is shown below. Note that computer output is in boldface, while our input is not. The categories defined by the graph in Figure 3.2 are entered in this example. Once the categories are set up, the first data entry of 4.0 gets fuzzified to the accomodative category. Note that the memberships are also presented in each category. The same value is entered again, and this time gets fuzzified to the very accomodative category. For the last data entry of 12.5, you see that only the very tight category holds membership for this value. In all cases you will note the memberships add up to 1.0.

```
> fuzzfier
```

Please type in a category name, e.g. Cool
Enter one word without spaces
When you are done, type 'done' :

```
v.accomodative
```

Type in the lowval, midval and highval
for each category, separated by spaces
e.g. 1.0 3.0 5.0 :

```
0 3 6
```

Please type in a category name, e.g. Cool
Enter one word without spaces
When you are done, type 'done' :

```
accomodative
```

Type in the lowval, midval and highval

for each category, separated by spaces
e.g. 1.0 3.0 5.0 :

```
3 6 9
```

Please type in a category name, e.g. Cool
Enter one word without spaces
When you are done, type 'done' :

```
tight
```

Type in the lowval, midval and highval
for each category, separated by spaces
e.g. 1.0 3.0 5.0 :

```
5 8.5 12
```

Please type in a category name, e.g. Cool
Enter one word without spaces
When you are done, type 'done' :

```
v.tight
```

Type in the lowval, midval and highval
for each category, separated by spaces
e.g. 1.0 3.0 5.0 :

```
10 12 14
```

Please type in a category name, e.g. Cool
Enter one word without spaces
When you are done, type 'done' :

```
done
```

```
====================================
==Fuzzifier is ready for data==
====================================
```

input a data value, type 0 to terminate
```
4.0
```

Output fuzzy category is ==> accomodative<==
category   membership
----
v.accomodative   0.666667
accomodative   0.333333
tight   0
v.tight   0

input a data value, type 0 to terminate

```
4.0
```

Output fuzzy category is ==> v.accomodative<==
category   membership
----
v.accomodative   0.666667
accomodative   0.333333
tight   0
v.tight   0

input a data value, type 0 to terminate
```
7.5
```

Output fuzzy category is ==> accomodative<==
category   membership
----
v.accomodative   0
accomodative   0.411765
tight   0.588235
v.tight   0

input a data value, type 0 to terminate
```
11.0
```

Output fuzzy category is ==> tight<==
category   membership
----
v.accomodative   0
accomodative   0
tight   0.363636
v.tight   0.636364

input a data value, type 0 to terminate
```
12.5
```

Output fuzzy category is ==> v.tight<==
category   membership
----
v.accomodative   0
accomodative   0
tight   0
v.tight   I

input a data value, type 0 to terminate
```
0
```

All done. Have a fuzzy day !

Fuzzy Associative Map

Fuzziness can also enter neural networks to define the weights from fuzzy sets. A comparison between *expert systems* and fuzzy systems is important to understand in the context of neural networks. Expert systems are based on crisp rules. Such crisp rules may not always be available. Expert systems have to consider an exhaustive set of possibilities. Such sets may not be known before hand. When crisp rules are not possible, and when it is not known if the possibilities are exhaustive, the expert systems approach is not a good one.

Some neural networks, through the features of training and learning, can function in the presence of unexpected situations. Therein neural networks have an advantage over expert systems, and they can manage with far less information than expert systems need.

Fuzziness in neural networks can sometimes be what is called *fuzzy cognitive maps*. Qualitative descriptors like *large negative, small positive*, and so on, can be translated into numbers similar to degree of membership in a set, and vice versa. In trying to parallel park a car against a curb, one may try to calculate precisely with the help of mathematics, the angle by which the steering wheel has to be turned at any instant during such an operation. On the other hand, one can consider fuzzy operations like, turning the steering wheel a little to the right or more to the left, and so on until the car gets parked in place. In an example like this, a neural network can hone in on the right value of the angle by which the steering wheel needs to be turned, through a process of unsupervised learning. Our treatment of fuzziness in neural networks is with the discussion of *Fuzzy Associative Memories*, abbreviated as FAMs, developed by Bart Kosko. This topic and the C++ implementation are in Chapter 9.

Fuzzy Systems and Membership Rules

So far we have considered how fuzzy logic plays a role in neural networks. The converse relationship, neural networks in fuzzy systems, is also an active area of research. In order to build a fuzzy system, you must have a set of membership rules for fuzzy categories. It is sometimes difficult to deduce these membership rules with a given set of complex data. Why not use a neural network to define the fuzzy rules for you? A neural network is good at discovering relationships and patterns in data, and can be used to preprocess data in a fuzzy system. Further, a neural network that can learn new relationships with new input data, can be used to refine fuzzy rules to create a fuzzy adaptive system. Neural trained fuzzy systems are being used in many commercial applications, especially in Japan:

◆ The Laboratory for International Fuzzy Enginnering Research (LIFE) in Yokohama, Japan has a backpropagation neural network that derives fuzzy rules and membership functions. The LIFE system has been successfully applied to a foreign-exchange trade support system with approximately 5000 fuzzy rules.

◆ Ford Motor Company has developed trainable fuzzy systems for automobile idle-speed control.

◆ National Semiconductor Corporation has a software product called NeuFuz that supports the generation of fuzzy rules with a neural network for control applications.

◆ A number of Japanese consumer and industrial products use neural networks with fuzzy systems, including vacuum cleaners, rice cookers, washing machines and photocopying machines.

Summary

In this chapter you read about fuzzy logic and fuzzy sets, and simple operations on fuzzy sets. Fuzzy logic unlike Boolean logic has more than two on or off categories to describe behavior of systems. You use membership values for data in fuzzy categories, which may overlap. In this chapter you also developed a fuzzifier program in C++ that takes crisp values and converts them to fuzzy values, based on categories and memberships that you define. For use with neural networks, fuzzy logic can serve as a postprocessing or pre-processing filter. Kosko developed neural networks that use fuzziness, and called them fuzzy associative memories, which will be discussed in later chapters. You also read about how neural networks can be used in fuzzy systems to define membership functions and fuzzy rules.

Chapter 4

Constructing a Neural Network

First Example for C++ Implementation

The neural network we presented in Chapter 1 is an example of a Hopfield network with a single layer. Let's say we place four neurons, all connected to one another on this layer, as shown previously in Figure 1.2. Some of these connections have a positive weight and the rest have a negative weight. Now we present a C++ implementation of this network. You may recall from the earlier presentation of this example, that we used two input patterns to determine the weight matrix. The network recalls them when the inputs are presented to the network, one at a time. These inputs are *binary* and *orthogonal* so that their recall is assured. Each component of a binary input pattern is either a 0 or a 1. Two vectors are orthogonal when their dot product—the sum of the products of their corresponding components—is zero. An example of a binary input pattern is 1 0 1 0 0. An example of a pair of orthogonal vectors is (0, 1, 0, 0, 1) and (1, 0, 0, 1, 0). An example of a pair of vectors which are not orthogonal is (0, 1, 0, 0, 1) and (1, 1, 0, 1, 0). These last two vectors have a dot product of 1, different from 0.

The two patterns we want the network to recall are A = (1, 0, 1, 0) and B = (0, 1, 0, 1). The weight matrix W is given below:

$$W = \begin{array}{cccc} 0 & -3 & 3 & -3 \\ -3 & 0 & -3 & 3 \\ 3 & -3 & 0 & -3 \\ -3 & 3 & -3 & 0 \end{array}$$

We need a threshold function also, and we define it using a threshold value, θ, as follows.

$$f(t) = \begin{cases} 1 & \text{if } t >= \theta \\ 0 & \text{if } t < \theta \end{cases}$$

The threshold value θ is used as a cut off value for the activation of a neuron to enable it to fire. The activation should equal or exceed the threshold value for the neuron to fire, meaning to have output 1. For our Hopfield network, θ is taken as 0. There are four neurons in the only layer in this network. The first node's output is the output of the **threshold** function. The argument for the **threshold** function is the activation of the node. And the activation of the node is the dot product of the input vector and the first column of the weight matrix. So if the input vector is A, the dot product becomes 3, and f(3) = 1. And the dot products of the second, third, and fourth nodes become –6, 3, and –6, respectively. The corresponding outputs therefore are 0, 1, and 0. This means that the output of the network is the vector (1, 0, 1, 0), which is the same as the input pattern. Therefore, the network has recalled the pattern as presented. When B is presented, the dot product obtained at the first node is –6 and the output is 0. The activations of all the four nodes together with the **threshold** function give (0, 1, 0, 1) as output from the network, which means that the network recalled B as well. The weight matrix worked well with both input patterns, and we do not need to modify it.

Classes in C++ Implementation

In our C++ implementation of this network, there are the following classes: a *network* class, and a *neuron* class. In our implementation, we create the network with four neurons, and these four neurons are all connected to one another. A neuron is not self-connected, though. That is, there is no edge in the directed graph representing the network, where the edge is from one node to itself. But for simplicity, we could pretend that such a connection exists carrying a weight of 0, so that the weight matrix has 0's in its principal diagonal.

The functions that determine the neuron activations and the network output are declared public. Therefore they are visible and accessible without restriction. The activations of the neurons are calculated with functions defined in the neuron class. When there are more than one layers in a neural network, the outputs of neurons in one layer become the inputs for neurons in the next layer. In order to facilitate passing the outputs from one layer as inputs to another layer, our C++ implementations compute the neuron outputs in the network class. For this reason the threshold function is made a member of the network class. We do this for the Hopfield network as well. To see if the network has achieved correct recall, you make comparisons between the presented pattern and the network output, component by component.

C++ Program for a Hopfield Network

Every C++ program has two components: One is the header file with all of the class declarations and lists of include library files; the other is the source file that includes the header file and the detailed descriptions of the member functions of the classes declared in the header file. You also put the function **main** in the source file. Most of the computations are done by class member functions, when class objects are created in the function **main**, and calls are made to the appropriate functions. The header file has a **.h** (or **.hpp**) extension, as you know, and the source file has a **.cpp** extension, to indicate that it is a C++ code file. It is possible to have the contents of the header file written at the beginning of the **.cpp** file and work with one file only, but separating the declarations and implementations into two files allows you to change the implementation of a class(**.cpp**) without changing the interface to the class (**.h**).

Header File for C++ Program for Hopfield Network

Listing 4.1 contains Hop.h, the header file for the C++ program for the Hopfield network. The include files listed in it are the stdio.h, iostream.h, and math.h. The iostream.h file contains the declarations and details of the C++ streams for input and output. A **network** class and a **neuron** class, are declared in Hop.h. The data members and member functions are declared within each class, and their accessibility is specified by the keywords *protected* or *public*.

Listing 4.1   Header file for C++ program for Hopfield network.

```
//Hop.h       V. Rao, H. Rao
//Single layer Hopfield Network with 4 neurons

#include <stdio.h>
#include <iostream.h>
#include <math.h>

class neuron
{
protected:
    int activation;
    friend class network;
public:
    int weightv[4];
    neuron() {};
    neuron(int *j) ;
    int act(int, int*);
};


class network
{
public:
    neuron   nrn[4];
    int output[4];
    int threshld(int) ;
    void activation(int j[4]);
    network(int*,int*,int*,int*);

};
```

Notes on the Header File Hop.h

Notice that the data item activation in the neuron class is declared as protected. In order to make the member activation of the neuron class accessible to the network class, the network is declared a *friend* class in the class neuron. Also, there are two constructors for the class neuron. One of them creates the object neuron without initializing any data members. The other creates the object neuron and initializes the connection weights. One of the connection weights is necessarily 0. It corresponds to a nonexistent recurrent connection at the neuron.

Source Code

Listing 4.2 contains the source code for the C++ program for a Hopfield network in the file Hop.cpp. The member functions of the classes declared in Hop.h are implemented here. The function main contains the input patterns, values to initialize the weight matrix, and calls to the constructor of network class and other member functions of the network class.

Listing 4.2   Source code for C++ program for Hopfield network.

```
//Hop.cpp V. Rao, H. Rao
//Single layer Hopfield Network with 4 neurons

#include "hop.h"

neuron::neuron(int *j)
{
int i;
for(i=0;i<4;i++)
    {
    weightv[i]= *(j+i);
    }

}

int neuron::act(int m, int *x)
{
int i;
int a=0;

for(i=0;i<m;i++)
    {
    a += x[i]*weightv[i];
```

```
        }
    return a;
    }

    int network::threshld(int k)
    {
    if(k>=0)
        return (1);
    else
        return (0);
    }

    network::network(int a[4],int b[4],int c[4],int d[4])
    {
    nrn[0] = neuron(a) ;
    nrn[1] = neuron(b) ;
    nrn[2] = neuron(c) ;
    nrn[3] = neuron(d) ;
    }

    void network::activation(int *patrn)
    {
    int i,j;

    for(i=0;i<4;i++)
        {
        for(j=0;j<4;j++)
            {
            cout<<"\n nrn["<<i<<"].weightv["<<j<<"] is "
                <<nrn[i].weightv[j];
            }
        nrn[i].activation = nrn[i].act(4,patrn);
        cout<<"\nactivation is "<<nrn[i].activation;
        output[i]=threshld(nrn[i].activation);
        cout<<"\noutput value is   "<<output[i]<<"\n";
        }
    }

    void main ()
    {
    int patrn1[]= {1,0,1,0},i;
    int wt1[]= {0,-3,3,-3};
    int wt2[]= {-3,0,-3,3};
    int wt3[]= {3,-3,0,-3};
    int wt4[]= {-3,3,-3,0};

    cout<<"\nTHIS PROGRAM IS FOR A HOPFIELD NETWORK WITH A SINGLE
        LAYER OF";
    cout<<"\n4 FULLY INTERCONNECTED NEURONS. THE NETWORK SHOULD
        RECALL THE";
    cout<<"\nPATTERNS 1010 AND 0101 CORRECTLY.\n";
```

```
//create the network by calling its constructor.
// the constructor calls neuron constructor as many times as
    the number of
// neurons in the network.
network h1(wt1,wt2,wt3,wt4);

//present a pattern to the network and get the activations of
    the neurons
h1.activation(patrn1);

//check if the pattern given is correctly recalled and give
    message
for(i=0;i<4;i++)
    {
    if (h1.output[i] == patrn1[i])
        cout<<"\n pattern= "<<patrn1[i]<<
        "  output = "<<h1.output[i]<<"  component matches";
    else
        cout<<"\n pattern= "<<patrn1[i]<<
        "  output = "<<h1.output[i]<<
        "  discrepancy occurred";
    }
cout<<"\n\n";
int patrn2[]= {0,1,0,1};
h1.activation(patrn2);
for(i=0;i<4;i++)
    {
    if (h1.output[i] == patrn2[i])
        cout<<"\n pattern= "<<patrn2[i]<<
        "  output = "<<h1.output[i]<<"  component matches";
    else
        cout<<"\n pattern= "<<patrn2[i]<<
        "  output = "<<h1.output[i]<<
        "  discrepancy occurred";
        }

}
```

Comments on the C++ Program for Hopfield Network

Note the use of the output stream operator **cout<<** to output text strings or numerical output. C++ has **istream** and **ostream** classes from which the **iostream** class is derived. The standard input and output streams are **cin** and **cout**, respectively, used, correspondingly, with the operators >> and <<.

Use of **cout** for the output stream is much simpler than the use of the C function **printf**. As you can see, there is no formatting suggested for output. However, there is a provision that allows you to format the output, while using **cout**.

Also note the way comments are introduced in the program. The line with comments should start with a double slash //. Unlike C, the comment does not have to end with a double slash. Of course, if the comments extend to subsequent lines, each such line should have a double slash at the start. You can still use the pair, /* at the beginning with */ at the end of lines of comments, as you do in C. If the comment continues through many lines, the C facility will be handier to delimit the comments.

The neurons in the network are members of the network class and are identified by the abbreviation **nrn**. The two patterns, 1010 and 0101, are presented to the network one at a time in the program.

Output from the C++ Program for Hopfield Network

The output from this program is as follows and is self-explanatory. Computer output is boldface. When you run this program, you're likely to see a lot of output whiz by, so in order to leisurely look at the output use redirection. Type **Hop > filename**, and your output will be stored in a file, which you can edit with any text editor or list by using the **type filename | more** command.

> **THIS PROGRAM IS FOR A HOPFIELD NETWORK WITH A SINGLE LAYER OF 4 FULLY INTERCONNECTED NEURONS. THE NETWORK SHOULD RECALL THE PATTERNS 1010 AND 0101 CORRECTLY.**
>
> **nrn[0].weightv[0] is 0**
> **nrn[0].weightv[1] is -3**
> **nrn[0].weightv[2] is 3**
> **nrn[0].weightv[3] is -3**
> **activation is 3**
> **output value is 1**
>
> **nrn[1].weightv[0] is -3**
> **nrn[1].weightv[1] is 0**
> **nrn[1].weightv[2] is -3**
> **nrn[1].weightv[3] is 3**
> **activation is -6**
> **output value is 0**

nrn[2].weightv[0] is 3
nrn[2].weightv[1] is -3
nrn[2].weightv[2] is 0
nrn[2].weightv[3] is -3
activation is 3
output value is 1

nrn[3].weightv[0] is -3
nrn[3].weightv[1] is 3
nrn[3].weightv[2] is -3
nrn[3].weightv[3] is 0
activation is -6
output value is 0

pattern= 1 output = 1 component matches
pattern= 0 output = 0 component matches
pattern= 1 output = 1 component matches
pattern= 0 output = 0 component matches

nrn[0].weightv[0] is 0
nrn[0].weightv[1] is -3
nrn[0].weightv[2] is 3
nrn[0].weightv[3] is -3
activation is -6
output value is 0

nrn[1].weightv[0] is -3
nrn[1].weightv[1] is 0
nrn[1].weightv[2] is -3
nrn[1].weightv[3] is 3
activation is 3
output value is 1

nrn[2].weightv[0] is 3
nrn[2].weightv[1] is -3
nrn[2].weightv[2] is 0
nrn[2].weightv[3] is -3
activation is -6
output value is 0

nrn[3].weightv[0] is -3
nrn[3].weightv[1] is 3
nrn[3].weightv[2] is -3
nrn[3].weightv[3] is 0
activation is 3
output value is 1

pattern= 0 output = 0 component matches
pattern= 1 output = 1 component matches

pattern= 0 output = 0 component matches
pattern= 1 output = 1 component matches

Further Comments on the Program and its Output

Let us recall our previous discussion of this example in Chapter 1. What does the network give as output if we present a pattern different from both A and B? If C = (0, 1, 0, 0) is the input pattern, the dot products would be –3, 0, –3, 3 making the outputs of the neurons 0 ,1,0,1, meaning that B would be recalled. This is quite interesting, because if we intended to input B, and we made a slight error and ended up presenting C instead, the network would recall B. You can run the program by changing the pattern to 0, 1, 0, 0 and compiling again, to see that the B pattern is recalled.

Another element about the example in Chapter 1 is that the weight matrix W is not the only weight matrix that would enable the network to recall the patterns A and B correctly. If we replace the 3 and –3 in the matrix with 2 and –2, respectively, the resulting matrix would facilitate the same performance from the network. One way for you to check this is to change the wt1, wt2, wt3, wt4 given in the program accordingly, and compile and run the program again. The reason why both of the weight matrices work is that they are closely related. In fact, one is a scalar (constant) multiple of the other, that is, if you multiply each element in the matrix by the same scalar, namely 2/3, you get the corresponding matrix in cases where 3 and –3 are replaced with 2 and –2, respectively.

A New Weight Matrix to Recall More Patterns

Let's continue to discuss this example. Suppose we are interested in having the patterns E = (1, 0, 0, 1) and F = (0, 1, 1, 0) also recalled correctly, in addition to the patterns A and B. In this case we would need to train the network and come up with a learning algorithm, which we will discuss in more detail later in the book. We come up with the matrix W1, which follows.

$$W_1 = \begin{array}{cccc} 0 & -5 & 4 & 4 \\ -5 & 0 & 4 & 4 \\ 4 & 4 & 0 & -5 \\ 4 & 4 & -5 & 0 \end{array}$$

Try to use this modification of the weight matrix in the source program, and then compile and run the program to see that the network successfully recalls all four patterns A, B, E, and F.

Weight Determination

You may be wondering about how these weight matrices were developed in the previous example, since so far we've only discussed how the network does its job, and how to implement the model. You have learned that the choice of weight matrix is not necessarily unique. But you want to be assured that there is some way, besides trial and error, in which to construct a weight matrix. You can go about this in the following way.

Binary to Bipolar Mapping

Let's look at the previous example again. You have seen that by replacing each 0 in a binary string with a –1, you get the corresponding bipolar string. If you keep all 1's the same, and replace each 0 with a –1, you will have a formula for the above option. You can use the following function to each bit in the string:

```
f(x) = 2x - 1
```

When you give the binary bit x, you get the corresponding bipolar character f(x)

NOTE

For inverse mapping, which turns a bipolar string into a binary string, you use the following function:

```
f(x) = (x + 1) / 2
```

When you give the bipolar character x, you get the corresponding binary bit f(x)

N O T E

Pattern's Contribution to Weight

Next, we work with the bipolar versions of the input patterns. You take the patterns to be recalled one at a time and determine its contribution to the weight matrix of the network. The contribution of each pattern is itself a matrix. The size of such a matrix is the same as the weight matrix of the network. Then add these contributions, in the way matrices are added, and you end up with the weight matrix for the network, which is also referred to as the *correlation* matrix. Let us find the contribution of the pattern A = (1, 0, 1, 0):

First, we notice that the binary to bipolar mapping of A = (1, 0, 1, 0) gives the vector (1, –1, 1, –1).

Then we take the transpose, and multiply, the way matrices are multiplied, and we see what is below.

```
1   [1   -1   1   -1]         1   -1   1   -1
1                     =      -1    1  -1    1
1                            1    -1   1   -1
1                            -1    1  -1    1
```

Now subtract 1 from each element in the main diagonal (that runs from top left to bottom right). This operation gives the same result as subtracting the identity matrix from the given matrix, obtaining 0's in the main diagonal. The resulting matrix, which is given next, is the contribution of the pattern (1, 0, 1, 0) to the weight matrix.

```
 0   -1    1   -1
-1    0   -1    1
 1   -1    0   -1
-1    1   -1    0
```

Similarly, we can calculate the contribution from the pattern B = (0, 1, 0, 1) by verifying that pattern B's contribution is the same matrix as pattern A's contribution. Therefore, the matrix of weights for this exercise is the matrix W shown here.

$$W = \begin{matrix} 0 & -2 & 2 & -2 \\ -2 & 0 & -2 & 2 \\ 2 & -2 & 0 & -2 \\ -2 & 2 & -2 & 0 \end{matrix}$$

Autoassociative Network

The Hopfield network just shown has the feature that the network associates an input pattern with itself in recall. This makes the network an *autoassociative* network. The patterns used for determining the proper weight matrix are also the ones that are autoassociatively recalled. These patterns are called the *exemplars*. A pattern other than an exemplar may or may not be recalled by the network. Of course, when you present the pattern 0 0 0 0, it is recalled, even though it is not an exemplar pattern.

Orthogonal Bit Patterns

You may be wondering how many patterns the network with four nodes is able to recall. Let us first consider how many different bit patterns are orthogonal to a given bit pattern. This question really refers to bit patterns in which at least one bit is equal to 1. A little reflection tells us that if two bit patterns are to be orthogonal, they cannot both have 1's in the same position, since the dot product would need to be 0. In other words, a bit-wise logical AND operation of the two bit patterns has to result in a 0. This suggests the following. If a pattern P has k, less than 4, bit positions with 0 (and so 4-k bit positions with 1), and if pattern Q is to be orthogonal to P, then Q can have 0 or 1 in those k positions, but it must have only 0 in the rest 4-k positions. Since there are two choices for each of the k positions, there are 2^k possible patterns orthogonal to P. This number 2^k of patterns includes the pattern with all zeroes. So there really are 2^k-1 non-zero patterns orthogonal to P. Some of these 2^k-1 patterns are not orthogonal to each other. As an example, P can be the pattern 0 1 0 0 which has k=3 positions with 0. There are $2^3-1=7$ non-zero patterns orthogonal to 0 1 0 0. Among these are patterns 1 0 1 0 and 1 0 0 1, which are not orthogonal to each other, since their dot product is 1and not 0.

Network Nodes and Input Patterns

Since our network has four neurons in it, it also has four nodes in the *directed graph* that represents the network. These are laterally connected because connections are established from node to node. They are lateral because the nodes are all in the same layer. We started with the patterns A = (1, 0, 1, 0) and B = (0, 1, 0, 1) as the exemplars. If we take any other nonzero pattern that is orthogonal to A, it will have a 1 in a position where B also has a 1. So the new pattern will not be orthogonal to B. Therefore, the orthogonal set of patterns that contains A and B can have only those two as its elements. If you remove B from the set, you can get (at most) two others to join A to form an orthogonal set. They are the patterns (0, 1, 0, 0) and (0, 0, 0, 1).

If you follow the procedure described earlier to get the correlation matrix, you will most likely get the following weight matrix:

$$
W = \begin{matrix}
0 & -1 & 3 & -1 \\
-1 & 0 & -1 & -1 \\
3 & -1 & 0 & -1 \\
-1 & -1 & -1 & 0
\end{matrix}
$$

With this matrix, pattern A is recalled, but the zero pattern (0, 0, 0, 0) is obtained for the two patterns (0, 1, 0, 0) and (0, 0, 0, 1). Once the zero pattern is obtained, its own recall will be stable.

Second Example for C++ Implementation

Recall the cash register game from the show *Price is Right* used as one of the examples in Chapter 1. This example led to the description of the Perceptron neural network. We will now resume our discussion of the Perceptron model and follow up with its C++ implementation. Keep the cash register game example in mind as you read the following C++ implementation of the Perceptron model. Also note that the input signals in this example are not binary, but they are real numbers. It is because the prices of the items the contestant has to choose are real numbers (dollars and cents). A Perceptron has one layer of input neurons and one layer of single or multiple output neurons. Each input layer neuron is connected to each neuron in the output layer.

C++ Implementation of Perceptron Network

In our C++ implementation of this network, we have the following classes: We have separate classes for input neurons and output neurons. The **ineuron** class is for the input neurons. This class has weight and activation as data members. The **oneuron** class is similar and is for the output neuron. It is declared as a **friend** class in the **ineuron** class.. The output neuron class has also a data member called *output*. There is a **network** class, which is a **friend** class in the **oneuron** class. An instance of the **network** class is created with four input neurons. These four neurons are all connected with one output neuron.

The member functions of the **ineuron** class are: (1) a default constructor, (2) a second constructor that takes a real number as an argument, and (3) a function that calculates the output of the input neuron. The constructor taking one argument uses that argument to set the value of the weight on the connection between the input neuron and the output neuron. The functions that determine the neuron activations and the network output are declared public. The activations of the neurons are calculated with functions defined in the neuron classes. A threshold value is used by a member function of the output neuron to determine if the neuron's activation is large enough for it to fire, giving an output of 1.

Header File

Listing 4.3 contains **percept.h**, the header file for the C++ program for the Perceptron network. percept.h contains the declarations for three classes, one for input neurons, one for output neurons, and one for network.

Listing 4.3   The percept.h header file.

```
//percept.h          V. Rao, H. Rao
// Perceptron model

#include <stdio.h>
#include <iostream.h>
#include <math.h>

class ineuron
{
protected:
    float weight;
```

```
        float activation;
        friend class oneuron;
    public:
        ineuron() {};
        ineuron(float j) ;
        float act(float x);
    };

    class oneuron
    {
    protected:
        int output;
        float activation;
        friend class network;
    public:
        oneuron() { };
        void actvtion(float x[4], ineuron *nrn);
        int outvalue(float j) ;
    };

    class network
    {
    public:
        ineuron    nrn[4];
        oneuron    onrn;
        network(float,float,float,float);

    };
```

Implementation of Functions

The network is designed to have four neurons in the input layer. Each of them is a class ineuron, and these are member classes in the class network. There is one explicitly defined output neuron of the class **oneuron**. The network constructor also invokes the neuron constructor for each input layer neuron in the network by providing it with the initial weight for its connection to the neuron in the output layer. The constructor for the output neuron is also invoked by the network constructor, at the same time initializing the output and activation data members of the output neuron each to zero. To make sure there is access to needed information and functions, the output neuron is declared a **friend** class in the class **ineuron**. The network is declared as a **friend** class in the class **oneuron**.

Source Code for Perceptron Network

Listing 4.4 contains the source code in percept.cpp for the C++ implementation of the Perceptron model previously discussed.

Listing 4.4   Source code for Perceptron model.

```
//percept.cpp V. Rao, H. Rao
//Perceptron model

#include "percept.h"
#include "stdio.h"
#include "stdlib.h"

ineuron::ineuron(float j)
{
weight= j;
}

float ineuron::act(float x)
{
float a;

a = x*weight;

return a;
}

void oneuron::actvtion(float *inputv, ineuron *nrn)
{
int i;
activation = 0;

for(i=0;i<4;i++)
    {
    cout<<"\nweight for neuron "<<i+1<<" is   "<<nrn[i].weight;
    nrn[i].activation = nrn[i].act(inputv[i]);
    cout<<"         activation is "<<nrn[i].activation;
    activation += nrn[i].activation;
    }

cout<<"\n\nactivation is   "<<activation<<"\n";
}

int oneuron::outvalue(float j)
{
if(activation>=j)
    {
    cout<<"\nthe output neuron activation \
exceeds the threshold value of "<<j<<"\n";
```

```
        output = 1;
        }
    else
        {
        cout<<"\nthe output neuron activation \
is smaller than the threshold value of "<<j<<"\n";
        output = 0;
        }

    cout<<" output value is "<< output;
    return (output);
}

network::network(float a,float b,float c,float d)
{
nrn[0] = ineuron(a) ;
nrn[1] = ineuron(b) ;
nrn[2] = ineuron(c) ;
nrn[3] = ineuron(d) ;
onrn = oneuron();
onrn.activation = 0;
onrn.output = 0;
}
void main (int argc, char * argv[])
{

float inputv1[]= {1.95,0.27,0.69,1.25};
float wtv1[]= {2,3,3,2}, wtv2[]= {3,0,6,2};
FILE * wfile, * infile;
int num=0, vecnum=0, i;
float threshold = 7.0;

if (argc < 2)
    {
    cerr << "Usage: percept Weightfile Inputfile";
    exit(1);
    }
// open   files

wfile= fopen(argv[1], "r");
infile= fopen(argv[2], "r");

if ((wfile == NULL) || (infile == NULL))
    {
    cout << " Can't open a file\n";
    exit(1);
    }

cout<<"\nTHIS PROGRAM IS FOR A PERCEPTRON NETWORK WITH AN
    INPUT LAYER OF";
cout<<"\n4 NEURONS, EACH CONNECTED TO THE OUTPUT NEURON.\n";
```

```
cout<<"\nTHIS EXAMPLE TAKES REAL NUMBERS AS INPUT SIGNALS\n";

//create the network by calling its constructor.
//the constructor calls neuron constructor as many times as
    the number of
//neurons in input layer of the network.

cout<<"please enter the number of weights/vectors \n";
cin >> vecnum;

for (i=1;i<=vecnum;i++)
    {
    fscanf(wfile,"%f %f %f %f\n",
    &wtv1[0],&wtv1[1],&wtv1[2],&wtv1[3]);
    network h1(wtv1[0],wtv1[1],wtv1[2],wtv1[3]);
    fscanf(infile,"%f %f %f %f \n",
        &inputv1[0],&inputv1[1],&inputv1[2],&inputv1[3]);
    cout<<"this is vector # " << i << "\n";
    cout << "please enter a threshold value, eg 7.0\n";
    cin >> threshold;

    h1.onrn.actvtion(inputv1, h1.nrn);
    h1.onrn.outvalue(threshold);
    cout<<"\n\n";
    }

fclose(wfile);
fclose(infile);
}
```

Comments on Your C++ Program

Notice the use of input stream operator **cin>>** in the C++ program, instead of the C function **scanf** in several places. The **iostream** class in C++ was discussed earlier in this chapter. The program works like this: First, the network input neurons are given their connection weights, and then an input vector is presented to their layer. A threshold value is specified, and the output neuron, does the weighted sum of its inputs, which are the outputs of the input layer neurons. This weighted sum is the activation of the output neuron, and it is compared with the threshold value, and the output neuron fires (output is 1) if the threshold value is not greater than its activation. It does not fire (output is 0) if its activation is smaller than the threshold value. In this implementation, neither supervised nor unsupervised training is incorporated.

Input/Output for percept.cpp

There are two data files used in this program. One is for setting up the weights, and the other for setting up the input vectors. On the command line, you enter the program name followed by the weight file name and the input file name. For this discussion (also on the accompanying disk for this book) create a file called weight.dat, which contains the following data:

```
2.0 3.0 3.0 2.0
3.0 0.0 6.0 2.0
```

These are two weight vectors. Create also an input file called input.dat with the two data vectors below:

```
1.95 0.27 0.69 1.25
0.30 1.05 0.75 0.19
```

During the execution of the program, you are first prompted for the number of vectors that are used (in this case, 2), then for a threshold value for the input/weight vectors (use 7.0 in both cases). You will then see the following output. **Note that the computer output is in boldface**.

> `> percept weight.dat input.dat`

THIS PROGRAM IS FOR A PERCEPTRON NETWORK WITH AN INPUT LAYER OF 4 NEURONS, EACH CONNECTED TO THE OUTPUT NEURON.

THIS EXAMPLE TAKES REAL NUMBERS AS INPUT SIGNALS
please enter the number of weights/vectors
2
this is vector # 1
please enter a threshold value, eg 7.0
7.0

weight for neuron 1 is 2	**activation is 3.9**
weight for neuron 2 is 3	**activation is 0.81**
weight for neuron 3 is 3	**activation is 2.07**
weight for neuron 4 is 2	**activation is 2.5**

activation is 9.28

the output neuron activation exceeds the threshold value of 7
 output value is 1

this is vector # 2

please enter a threshold value, eg 7.0
7.0

weight for neuron 1 is 3	**activation is 0.9**
weight for neuron 2 is 0	**activation is 0**
weight for neuron 3 is 6	**activation is 4.5**
weight for neuron 4 is 2	**activation is 0.38**

activation is 5.78

**the output neuron activation is smaller than the threshold value of 7
output value is 0**

Finally, try adding a data vector of (1.4, 0.6, 0.35, 0.99) to the data file. Add a weight vector of (2, 6, 8, 3) to the weight file and use a threshold value of 8.25 to see the result. You can use other values to experiment also.

Stability and Plasticity

A neural network operation is usually done in an iterative way, meaning that the procedure is repeated a certain number of times. These iterations are referred to as *cycles*. After each cycle, the input used may remain the same or change, or the weights may remain the same or change. Such change is based on the output of your completed cycle. If the number of cycles is not preset, and the network is allowed to go through cycles until some other criterion is met, the question of whether or not the termination of the iterative process occurs eventually, arises naturally.

Stability for a Neural Network

Stability refers to such convergence that facilitates an end to the iterative process. For example, if any two consecutive cycles result in the same output for the network, then there is no need to do more iterations. In this case, convergence has occurred, and the network has stabilized in its operation. If weights are being modified after each cycle, then convergence of weights would constitute stability for the network.

In some situations it takes many more iterations than you desire, to have output in two consecutive cycles to be the same. Then a tolerance level on the convergence criterion can be used. With a tolerance level, you accomplish early but satisfactory termination of the operation of the network.

Plasticity for a Neural Network

Suppose a network is trained to learn some patterns, and in this process the weights are adjusted according to an algorithm. After learning these patterns and encountering a new pattern, the network may modify the weights in order to learn the new pattern. But what if the new weight structure is not responsive to the new pattern? Then the network does not possess *plasticity*—the ability to deal satisfactorily with new *short-term* memory while retaining *long-term* memory. Attempts to endow a network with plasticity may have some adverse effects on the stability of your network.

Short-Term and Long-Term Memory

We alluded to short-term and long-term memory in the previous paragraph. Short-term memory is basically the information that is currently and perhaps temporarily being processed. It is manifested in the patterns that the network encounters. Long-term memory, on the other hand, is information that is already stored and is not being currently processed. As stated above, in a neural network, memory is usually characterized by patterns, and its storage is characterized by the weights used on the connections. The weights determine how an input is processed in the network to yield output. During the cycles of operation of a network, the weights may change. After convergence, they represent LTM, the long-term memory, as the weight levels achieved are stable.

Summary

You saw in this chapter, the C++ implementations of a simple Hopfield network and of a simple Perceptron network. What have not been included in them is an automatic iteration and a learning algorithm. For the examples used in this chapter to show C++ implementation, they were not necessary, and the emphasis was on the method of implementation. In a later chapter, you will read about the learning algorithms and examples of how to implement some of them.

You also read about the following concepts: Stability, Plasticity, Short-term memory, and Long-term memory. Much more can be said about them, in terms of the so-called noise-saturation dilemma, or stability–plasticity dilemma and what research has developed to address them. You can refer to one of the books in the Bibliography, if you are interested. As long as you are planning to use a neural network, which has been well understood for its purpose and construction, you will benefit from reading what we present here.

Chapter
5

Interactions: A Survey of Neural Network Models

Interactions

nteractions between neurons are depicted by their connection weight strengths. These connections can be for excitation or inhibition. It can be in a cooperative setting or a competitive setting. The network architecture suggests what kind of interactions will be present in the neural network. The following sections discuss these aspects as part of the discussion of network models.

Neural Network Models

You have been introduced in the preceding pages to the Perceptron model and the Hopfield network. You have learned that the differences between the models lie in their architecture, encoding, and recall. We aim now to give you a comprehensive picture of these and other neural network models. We will show details and implementations of some networks in later chapters.

The models we touch upon are the Perceptron, Hopfield, Adaline, Backpropagation, Bidirectional Associative Memory, Brain-State-in-a-Box, Neocognitron, Fuzzy Associative Memory, Fuzzy Cognitive Map, ART1, and ART2. C++ implementations of some of these and the role of fuzzy logic in some will be treated in the subsequent chapters. For now, our discussion will be about the components of a neural network. We will follow it with the description of some of the models.

Layers in a Neural Network

A neural network has its neurons divided into subgroups, or fields, and elements in each subgroup are placed in a row, or a column, in the diagram depicting the network. Each subgroup is then referred to as a layer of neurons in the network. A great many of the models of neural networks have two layers, quite a few have one layer, and some have three or more layers. A number of additional, so-called hidden layers are possible in some networks, such as the Backpropagation network. When the network has a single layer, the input signals are received at that layer, processing is done by its neurons, and output is generated at that layer. When more than one layer is present, the first field is for the neurons that supply the input signals for the neurons in the next layer.

Every network has a layer of input neurons, but in most of the networks, the sole purpose of these neurons is to feed the input to the next layer of neurons. However, there are feedback connections, or recurrent connections in some networks, so that the neurons in the input layer may also do some processing. In the Hopfield network you saw earlier, the input layer and output layer are the same. If any layer is present between the input and output layers, it may be referred to as a hidden layer in general, or as a layer with a special name taken after the researcher who proposed its inclusion to achieve certain performance from the network. Examples are the Grossberg and the Kohonen layers. The number of hidden layers is not limited except by the scope of the problem being addressed by the neural network.

A layer is also referred to as a field. Then the different layers can be designated as field A, field B, and so on, or shortly, F_A, F_B, and so on.

Single-Layer Network

A neural network with a single layer is also capable of processing for some important applications, such as integrated circuit implementations or assembly line control. The most common capability of the different models of neural networks is pattern recognition. But one network, called the Brain-State-in-a-Box, which is a single-layer neural network, can do pattern completion. Adaline is a network with A and B fields of neurons, but aggregation of input signals is done only by the field B neurons.

A single-layer neural network is exemplified by a simple Perceptron and also by a Hopfield network. While the Perceptron aims to evaluate a function that can take only one of two values for a given set of arguments (inputs), the Hopfield network makes the association between patterns. If the Hopfield network associates a pattern with itself, you have an example of autoassociation. Or you may characterize this as being able to recognize a given pattern. The idea of viewing it as a case of pattern recognition becomes more relevant if a pattern is presented with some *noise*, meaning that there is some slight deformation in the pattern, and if the network is able to relate it to the correct pattern.

The Perceptron has two layers, the second layer being the custodian of the output neuron, and the first holding the neurons that receive input(s). Also, the neurons in the same layer, the input layer in this case, are not interconnected, that is, no connections are made between two neurons in that same layer. On the other hand, in the Hopfield network, there is no separate output layer, and hence, it is strictly a single-layer network. In addition, the neurons are all fully connected with one another.

The ability of a Perceptron in evaluating functions was brought into question when Minsky and Papert proved that a simple function like **XOR** (the logical function **exclusive or**) could not be correctly evaluated by a Perceptron. If you apply this function to two logical variables that can have only 0, or *false*, and 1, or *true* as their values, the **XOR** function has a value of 1 if just one of the variables has value 1. If both variables are 0 or both variables are 1, then the **XOR** function has value 0. Minsky and Papert showed that it is impossible to come up with the proper set of weights for the neurons in the single layer of a simple Perceptron to evaluate the **XOR** function. The reason for this, later studies showed, is that such a Perceptron, one with a single layer of neurons, requires the domain of the function to be evaluated, to be *linearly separable*, by means of the values of the function. The concept of linear separability is explained next.

Linear Separability

What linearly separable means is, that a type of a linear barrier or a separator—a line in the plane, or a plane in the three-dimensional space, or a *hyperplane* in higher dimensions—should exist, so that the set of inputs that give rise to one value for the function all lie on one side of this barrier, while on the other side lie the inputs that do not yield that value for the function. A hyperplane is a surface in a higher dimension, but with a linear equation defining it much the same way a line in the plane and a plane in the three-dimensional space are defined.

To make the concept a little bit clearer, consider a problem similar to the XOR problem. Imagine a cube of 1-unit length for each of its edges and lying in the positive octant in a xyz rectangular coordinate system with one corner at the origin. The other corners or vertices are at points with coordinates (0, 0, 1), (0, 1, 0), (0, 1, 1), (1, 0, 0), (1, 1, 0), (1, 0, 1), and (1, 1, 1). Call the origin O, and the seven points listed as A, B, C, D, F, E, and G, respectively. Then any two faces opposite to each other are linearly separable because you can define the separating plane as the plane-half way between these two faces and also parallel to these two faces.

For example, consider the faces defined by the set of points O, A, B, and C and by the set of points D, E, F, and G. They are parallel and 1 unit apart, as you can see in Figure 5.1. The separating plane for these two faces can be seen to be one of many possible planes—any plane in between them and parallel to them. One example, for simplicity, is the plane that passes through the points (1/2, 0, 0), (1/2, 0, 1), (1/2, 1, 0), and (1/2, 1, 1). Of course, you need only specify three of those four points because a plane is uniquely determined by three points that are not all on the same line. So if the first set of points correspond to a value of say, +1 for the function, and the second set to a value of –1, then a single-layer Perceptron can determine the correct weights for the connections, even if you start with the weights being initially all 0.

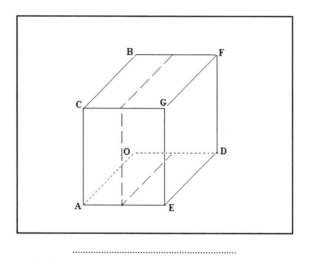

FIGURE 5.1 *Separating plane.*

Consider the set of points O, A, F, and G. This set of points cannot be linearly separated from the other vertices of the cube. In this case, it would be impossible for the single-layer Perceptron to determine the proper weights for the neurons in evaluating the type of function we have been discussing.

A Second Look at the XOR Function

By introducing a hidden layer of neurons in the Perceptron network and having connections between this and the output layers, the **XOR** function can be evaluated. The absence of the separability that we talked about earlier is overcome by having a second stage, so to speak, of connection weights.

Example of the Cube Revisited

Let us return to the example of the cube with vertices at the origin O, and the points labeled A, B, C, D, E, F, and G. Suppose the set of vertices O, A, F, and G give a value of 1 for the function to be evaluated, and the other vertices a –1. The two sets are not linearly separable as mentioned before. A simple Perceptron cannot evaluate this function.

Can the addition of another layer of neurons help? The answer is yes. What would be the role of this additional layer? The answer is that it will participate in the final processing for the problem after the previous layer has done some preprocessing. This can do two separations in the sense that the set of eight vertices can be separated—or partitioned—into three separable subsets. If this partitioning can also help collect within each subset, like vertices, meaning those that map onto the same value for the function, the network will succeed in its task of evaluating the function when the aggregation and thresholding is done at the output neuron.

Strategy

So the strategy is first to consider the set of vertices that give a value of +1 for the function and determine the minimum number of subsets that can be identified to be each separable from the rest of the vertices. It is evident that since the vertices O and A lie on one edge of the cube, they can form one subset that is separable. The other two vertices, viz., F and G, which correspond to the value +1 for the function, can form a second subset that is separable, too. We need not bother with the last four vertices from the point of view of further partitioning that subset. It is clear that one new layer of three neurons, one of which fires for the inputs corresponding to the vertices O and A, one for F, and G, and the third for the rest, will then facilitate the correct evaluation of the function at the output neuron.

Details

Table 5.1 lists the vertices and their coordinates, together with a flag that indicates to which subset in the partitioning the vertex belongs.

TABLE 5.1 *Partitioning of vertices of a cube.*

Vertex	Coordinates	Subset
O	(0, 0, 0)	1
A	(0, 0, 1)	1
B	(0, 1, 0)	2
C	(0, 1, 1)	2
D	(1, 0, 0)	2
E	(1, 0, 1)	2
F	(1, 1, 0)	3
G	(1, 1, 1)	3

The network, which is a two-layer Perceptron, has three neurons in the first layer and one output neuron in the second layer. Remember that we are counting those layers in which the neurons do the aggregation of the signals coming into them using the connection weights. The first layer with the three neurons is what is generally described as the hidden layer, since the second layer is not hidden and is at the extreme right in the layout of the neural network. Table 5.2 gives an example of the weights you can use for the connections between the input neurons and the hidden layer neurons. There are three input neurons, one for each coordinate of the vertex of the cube.

TABLE 5.2 *Weights for connections between input neurons and hidden layer neurons.*

Input Neuron#	Hidden Layer Neuron#	Connection Weight
1	1	1
1	2	0.1
1	3	-1
2	1	1
2	2	-1

Continued

Input Neuron#	Hidden Layer Neuron#	Connection Weight
2	3	-1
3	1	0.2
3	2	0.3
3	3	0.6

Now we give, in Table 5.3, the weights for the connections between the three hidden-layer neurons and the output neuron.

TABLE 5.3   *Weights for connection between the hidden-layer neurons and the output neuron.*

Hidden Layer Neuron#	Connection Weight
1	0.6
2	0.3
3	0.6

You cannot already see whether or not these weights will do the job. To determine the activations of the hidden-layer neurons, you need these weights, and you also need the threshold value at each neuron that does processing. A hidden-layer neuron will fire, that is, will output a 1, if the weighted sum of the signals it receives is greater than the threshold value. If the output neuron fires, the function value is taken as +1, and if it does not fire, the function value is –1. Table 5.4 gives the threshold values.

TABLE 5.4   *Threshold values.*

Layer	Neuron	Threshold Value
hidden	1	1.8
hidden	2	0.05
hidden	3	-0.2
output	1	0.5

Performance of the Perceptron

When you input the coordinates of the vertex G, which has 1 for each coordinate, the first hidden-layer neuron aggregates these inputs and gets a value of 2.2. Since 2.2 is more than the threshold value of the first neuron in the hidden layer, that neuron fires, and its output of 1 becomes an input to the output neuron on the connection with weight 0.6. But you need the activations of the other hidden-layer neurons as well. Let us describe the performance with coordinates of G as the inputs to the network. Table 5.5 describes this.

TABLE 5.5 *Results with coordinates of vertex G as input.*

Vertex/ Coordinates	Hidden Layer	Weighted Sum	Comment	Activation	Contribution To Output	Sum
G : 1, 1, 1	1	2.2	>1.8	1	0.6	
	2	-0.8	<0.05	0	0	
	3	-1.4	<-0.2	0	0	0.6

The weighted sum at the output neuron is 0.6, and it is greater than the threshold value 0.5. Therefore, the output neuron fires, and at the vertex G, the function is evaluated to have a value of +1.

Table 5.6 shows the performance of the network with the rest of the vertices of the cube. You will notice that the network computes a value of +1 at the vertices, O, A, F, and G, and a –1 at the rest.

TABLE 5.6 *Results with other inputs.*

Vertex/ Coordinates	Hidden Layer	Weighted Sum	Comment	Activation	Contribution To Output	Sum
O :0, 0, 0	1	0	<1.8	0	0	
	2	0	<0.05	0	0	
	3	0	>-0.2	1	0.6	0.6 *
						Continued

Vertex/ Coordinates	Hidden Layer	Weighted Sum	Comment	Activation	Contribution To Output	Sum
A :0, 0, I	I	0.2	<1.8	0	0	
	2	0.3	>0.05	I	0.3	
	3	0.6	>-0.2	I	0.6	0.9*
B :0, I, 0	I	I	<1.8	0	0	
	2	-I	<0.05	0	0	
	3	-I	<-0.2	0	0	0
C :0, I, I	I	1.2	<1.8	0	0	
	2	0.2	>0.05	I	0.3	
	3	-0.4	<-0.2	0	0	0.3
D :I, 0, 0	I	I	<1.8	0	0	
	2	.I	>0.05	I	0.3	
	3	-I	<-0.2	0	0	0.3
E :I, 0, I	I	1.2	<1.8	0	0	
	2	0.4	>0.05	I	0.3	
	3	-0.4	<-0.2	0	0	0.3
F :I, I, 0	I	2	>1.8	I	0.6	
	2	-0.9	<0.05	0	0	
	3	-2	<-0.2	0	0	0.6*

The output neuron fires as this value is greater than 0.5, the threshold value. The function value is +1.

Two-layer Networks

Many important neural network models have two layers. *Backpropagation* network, in its simplest form, is one example. Grossberg and Carpenter's ART1 paradigm uses a two-layer network. The *Counterpropagation* network has a Kohonen layer followed by a Grossberg layer. *Bidirectional Associative Memory*, (BAM), *Boltzman Machine, Fuzzy Associative Memory*, and *Temporal Associative Memory* are other two-layer networks. For autoassociation, a single-layer network could do the job, but for heteroassociation or other such mappings, you need at least a two-layer network.

Multilayer Networks

Kunihiko Fukushima's *Neocognitron*, noted for recognizing handwritten characters, is an example of a network with several layers. Some previously mentioned networks can also be multilayered from the addition of more hidden layers. It is also possible to combine two or more neural networks into one network, by creating appropriate connections between layers of one subnetwork to those of the others. This would certainly create a multi-layer network, overall.

Connections Between Layers

You have already seen some difference in the way connections are made between neurons in a neural network. In the Hopfield network, every neuron was connected to every other in the one layer that was present in the network. In the Perceptron, neurons within the same layer were not connected with one another, but the connections were between the neurons in one layer and those in the next layer. In the former case, the connections are described as being *lateral*. In the latter case the connections are *forward*, and the signals are fed forward within the network.

Two other possibilities also exist. All the neurons in any layer may have extra connections, with each neuron connected to itself. Such a connection is described as *recurrent*. The second possibility is that there are connections from the neurons in one layer to the neurons in the previous layer, in which case there is both forward and backward signal feeding. This occurs, if *feedback* is a feature for the network model. The type of layout for the network neurons and the type of connections between the neurons constitute the architecture of the particular model of the neural network.

Instar and Outstar

Outstar and *instar* are terms defined by Stephen Grossberg for ways of looking at neurons in a network. A neuron in a web of other neurons receives a large number of inputs from outside its boundaries. This is like a inwardly radiating star, hence, the term instar. Also, a neuron may be sending its output to many other destinations in the network. In this way it is acting as an outstar. Every neuron is thus simultaneously both an instar and an outstar. As an instar it receives stimuli from other parts of the network or from outside of the network. Note that the neurons in the input layer of a network primarily have connections away from them to the neurons in the next layer, and thus behave mostly as outstars. Neurons in the output layer have many connections coming to it and thus behave mostly as instars. A neural network performs its work through the constant interaction of instars and outstars.

Weights on Connections

Weight assignments on connections between neurons not only indicate the strength of the signal that is being fed for aggregation, but also the type of interaction between the two neurons. The type of interaction is one of cooperation or of competition. The cooperative type is suggested by a positive weight, and the competition by a negative weight, on the connection. The positive weight connection is meant for what is called *excitation*, while the negative weight connection is termed an *inhibition*.

Initialization of Weights

Initializing the network weight structure is part of what is called the *encoding phase* of a network operation. The encoding algorithms are several, differing by model and by application. You may have gotten the impression that the weight matrices used in the examples discussed in detail thus far have been arbitrarily determined; or if there is a method of setting them up, you are not told what it is.

It is possible to start with randomly chosen values for the weights and to let the weights be adjusted appropriately as the network is run through successive iterations. This would make it easier also. For example, under supervised training, if the error between the desired and computed output is used as a criterion in adjusting weights, one may as well set the initial weights to zero and let the training process take care of the rest. The small example that follows illustrates this point.

Small Example

Suppose you have a network with two input neurons and one output neuron, with forward connections between the input neurons and the output neuron, as shown in Figure 5.2. The network is required to output a 1 for the input patterns (1, 0) and (1, 1), and the value 0 for (0, 1) and (0, 0). There are only two connection weights w_1 and w_2.

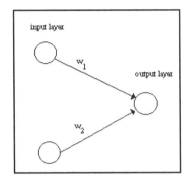

FIGURE 5.2 *Neural network with forward connections.*

Let us set initially both weights to 0. You need a **threshold** function also. Let us use the following threshold function, which is slightly different from the one used in a previous example:

```
f(x) = 1 if x > 0, and
f(x) = 0 if x ≤ 0
```

Now we need to know by what procedure we adjust the weights. The procedure we would apply for this example is as follows.

◆ If the output with input pattern (a, b) is as desired, then do not adjust the weights.

◆ If the output with input pattern (a, b) is smaller than what it should be, then increment each of w_1 and w_2 by 1.

◆ If the output with input pattern (a, b) is greater than what it should be, then subtract 1 from w_1 if the product aw_1 is smaller than 1, and adjust w_2 similarly.

Table 5.7 shows what takes place when we follow these procedures, and at what values the weights settle.

TABLE 5.7 *Adjustment of weights.*

Step	w₁	w₂	a	b	Activation	Output	Comment
1	0	0	1	1	0	0	desired output is 1. Increment both w's
2	1	1	1	1	2	1	output is what it should be
3	1	1	1	0	1	1	output is what it should be
4	1	1	0	1	1	1	output is 1. It should be 0.
5							Subtract 1 from w2
6	1	0	0	1	0	0	output is what it should be
7	1	0	0	0	0	0	output is what it should be
8	1	0	1	1	1	1	output is what it should be
9	1	0	1	0	1	1	output is what it should be

Table 5.7 shows that the network weight vector changed from an initial vector (0, 0) to the final weight vector (1, 0) in eight iterations. This example is not of a network for pattern matching. If you think about it, you will realize that the network is designed to fire if the first digit in the pattern is a 1 and not otherwise. An analogy for this kind of a problem is determining if a given image contains a specific object in a specific part of the image, such as a dot should occur in the letter i.

If the initial weights are chosen somewhat prudently and to make some particular relevance, then the speed of operation can be increased in the sense of convergence being achieved with fewer iterations than otherwise. Thus, encoding algorithms are important. We now present some of the encoding algorithms.

Initializing Weights for Autoassociative Networks

Consider a network that is to associate each input pattern with itself and which gets binary patterns as inputs. Make a bipolar mapping on the input pattern. That is, replace each 0 by –1. Call the mapped pattern the vector **x**, when written as a column vector. The transpose, the same vector written as a row vector, is **x**^{T}. You will get a matrix of order the size of **x** when you form the product **xx**^{T}. Obtain similar matrices for the other patterns you want the network to store. Add these matrices to give you the matrix of weights to be used initially. This process can be described with the following equation:

$$W = \Sigma_i \; x_i x_i{}^T$$

Weight Initialization for Heteroassociative Networks

Consider a network that is to associate one input pattern with another pattern and which gets binary patterns as inputs. Make a bipolar mapping on the input pattern. That is, replace each 0 by –1. Call the mapped pattern the vector **x** when written as a column vector. Get a similar bipolar mapping for the corresponding associated pattern. Call it **y**. You will get a matrix of size **x** by size **y**, when you form the product **xy**^{T}. Obtain similar matrices for the other patterns you want the network to store. Add these matrices to give you the matrix of weights to be used initially. The following equation restates this process:

$$W = \Sigma_i \; x_i y_i{}^T$$

On Center, Off Surround

In one of the many interesting paradigms you encounter in neural network models and theory, is the strategy of *winner takes all*. Well, if there should be one winner emerging from a crowd of neurons in a particular layer, there needs to be competition. Since everybody is for himself in such a competition, in this case every neuron for itself, it would be necessary to have lateral connections that indicate this circumstance. The lateral connections from any neuron to the others should have a negative weight. Or, the neuron with the highest activation is considered the winner and only its weights are modified in the training process, leaving the weights of others the same. Winner takes all means that only one neuron in that layer fires and the others do not. This can happen in a hidden layer or in the output layer.

In another situation, when a particular category of input is to be identified from among several groups of inputs, there has to be a subset of the neurons that are dedicated to seeing it happen. In this case, inhibition perks up for distant neurons, while excitation increases for the neighboring ones, as far as such a subset of neurons is concerned. You come across the phrase, *on center, off surround* to describe this phenomenon.

Weights also are the prime components in a neural network, as they reflect on the one hand the memory stored by the network, and on the other hand the basis for learning, and training.

Inputs

You have seen that mutually orthogonal patterns are required as inputs to the Hopfield network, which we discussed before for pattern matching. Similar restrictions on inputs are found also with some other neural networks. Sometimes it is not a restriction, but the purpose of the model makes a certain type of input natural. Certainly, in the context of pattern matching, binary input patterns make it simpler. Binary, bipolar, and analog signals are the varieties of inputs. Networks that accept analog signals as inputs are for continuous models, and those that require binary or bipolar inputs are for discrete models. Binary inputs can be fed to networks for continuous models, but analog signals cannot be input to networks for discrete models. With input possibilities being discrete or analog, and the model possibilities being discrete or continuous, there are potentially four situations, but one of them where analog inputs are considered for a discrete model is untenable.

An example of a continuous model is where a network is to adjust the angle, by which the steering wheel of a truck is to be turned, to back up the truck into a parking space. If a network is supposed to recognize characters of the alphabet, a means of discretization of a character allows the use of a discrete model.

What are the types of inputs for problems like image processing, handwriting analysis, and so on? Remembering that artificial neurons, as processing elements, do aggregation of their inputs by using connection weights, and that the output neuron uses a **threshold** function, you know that the inputs have to be numerical. A hand-written character can be superimposed on a grid, and the input can consist of the cells in each row of the grid, where a part of the character is present. In other words, the input corresponding to one character will be a set of binary sequences containing one sequence for each row of the grid. A 1 in a particular position in the sequence for a row shows that the character is present in that part of the grid, while 0 shows it is not. The size of the grid has to be big enough to accommodate the largest character under study.

Outputs

The output from some neural networks is in the form of a spatial pattern that can include a bit pattern, in some a binary function value, and in some others some analog signal. The type of mapping intended for the inputs determines the type of outputs, naturally. The output could be one of classifying the input data, or finding associations between patterns of the same dimension as the input.

The **threshold** functions do the final mapping of the activations of the output neurons into the network outputs. But the outputs from a single cycle of operation of a neural network may not be the final outputs, since you would iterate the network into further cycles of operation until you see convergence. If convergence looks like a possible eventuality, but is taking an awful lot of time and effort, that is, if it is too slow, you may assign a tolerance level and settle for the network to achieve near convergence. You would expect stability maintained in the process of achieving convergence.

Threshold Functions

The output of any neuron is the result of thresholding, if any, of its activation, which, in turn, is the weighted sum of the neuron's inputs. Thresholding sometimes is done for the sake of scaling down the activation and mapping it into a meaningful output for the problem, and sometimes for adding a bias, or for *squashing*, and so on. The most often-used **threshold** function is the **sigmoid** function. A **step** function, or a **ramp** function, or just a **linear** function can be used, as when you simply add the bias to the activation. **Sigmoid** function accomplishes mapping the activation into the interval [0, 1]. The equations are given as follows for the different **threshold** functions just mentioned.

Sigmoid Function

More than one function goes by the name, *sigmoid* function. They differ in their formulas and in their ranges. They all have a graph similar to a stretched letter **s**. We give below two such functions. The first is the hyperbolic tangent function with values in (–1, 1). The second is the logistic function and has values between 0 and 1. The graph of the logistic function is given in Fig. 5.3.

> 1. $f(x) = \tanh(x) = (e^x - e^{-x}) / (e^x + e^{-x})$
> 2. $f(x) = 1 / (1+ e^{-x})$

Figure 5.3 is the graph of the **sigmoid logistic** function (number 2, above).

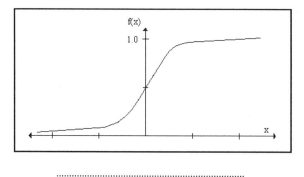

FIGURE 5.3   *The **sigmoid** function.*

Step Function

The **step** function is also frequently used as a **threshold** function. The function is 0 to start with and remains so to the left of some **threshold** value θ. A jump to 1 occurs for the value of the function to the right of θ, and the function remains at the level 1 from then on. In general, a **step** function can have a finite number of points at which jumps of equal or unequal size occur. When the jumps are equal and at many points, the graph will resemble a staircase. We are interested in a **step** function that goes from 0 to 1 in one step, as soon as the argument exceeds the **threshold** value θ. You could also have two values other than 0 and 1 in defining the range of values of such a **step** function. A graph of the **step** function follows in Figure 5.4.

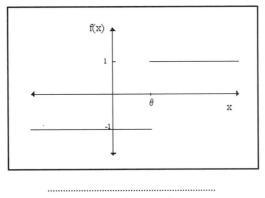

FIGURE 5.4   *The **step** function.*

Ramp Function

To describe the **ramp** function simply, first consider a **step** function that makes a jump from 0 to 1 at some point. Instead of letting it take a sudden jump like that at one point, let it gradually gain in value, along a straight line (looks like a ramp), over a finite interval reaching from an initial 0 to a final 1. Thus, you get a **ramp** function. The graph of a **ramp** function is illustrated in Figure 5.5.

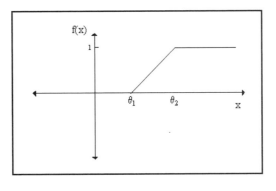

FIGURE 5.5   *Graph of a **ramp** function.*

Linear Function

A **linear** function is a simple one given by an equation of the form:

f(x) = αx + β

When α = 1, the application of this **threshold** function would amount to simply adding a bias equal to β to the sum of the inputs.

Applications

As briefly indicated before, the areas of application generally include auto- and heteroassociation, optimization, pattern recognition, data compression, data completion, signal filtering, image processing, and handwriting analysis. The type of connections in the network, and the type of learning algorithm used point to the type of application. For example, a network with lateral connections can do autoassociation, while a vector matching type can do optimization.

Some Neural Network Models

Adaline and Madaline

Adaline is the acronym for *adaptive linear element,* due to Bernard Widrow and Marcian Hoff. It is similar to a Perceptron. Inputs are real numbers in the interval [–1,+1], and learning is based on the criterion of minimizing the average squared error. Adaline has a high capacity to store patterns. Madaline stands for many Adalines, and is a neural network that is widely used. It is composed of field A and field B neurons, and there is one connection from each field A neuron to each field B neuron. Figure 5.6 shows a diagram of the Madaline.

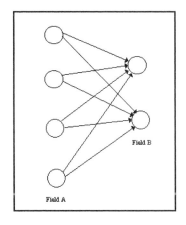

FIGURE 5.6   *The Madaline model.*

Backpropagation

Backpropagation paradigm and *algorithm* are developed by Paul Werbos. This type of network is the most common in use, due to its ease of training. It is estimated that over 80% of all neural network projects in development use backpropagation. In backpropagation, there are two phases in its learning cycle, one to propagate the input pattern and the other to adapt the output. It is the error signals that are backpropagated in the network operation to the hidden layer(s). The portion of the error signal that a hidden-layer neuron receives in this process is in relation to the contribution of the layer to the output. Adjusting on this basis the weights of the connections, the squared error is reduced in each cycle and finally minimized. Some other appropriate function to be minimized through weight modification could be used instead of the squared error.

Figure for Backpropagation Network

You will find in Figure 7.1 in Chapter 7, the layout of the nodes that represent the neurons in a Backpropagation network and the connections between them. For now, you try your hand at drawing this layout based on the following description, and compare your drawing with Figure 7.1. There are three fields of neurons. The connections are forward, and are from each neuron in a layer to every neuron in the next layer. There are no lateral or recurrent connections. Labels on connections indicate weights. Keep in mind that the number of neurons is not necessarily the same in different layers, and this fact should be evident in your choice of notation for the weights.

Bidirectional Associative Memory

Bidirectional Associative Memory, (BAM), developed by Bart Kosko, is a network with feedback connections from the output layer to the input layer. It associates a member of the set of input patterns with a member of the set of output patterns that is the closest, and thus it does heteroassociation. The patterns can be with binary or bipolar values. If all possible input patterns are known, the matrix of connection weights can be determined as the sum of matrices obtained by taking the matrix product of an input vector (as a column vector) with its transpose (written as a row vector).

The pattern obtained from the output layer in one cycle of operation is fed back at the input layer at the start of the next cycle. The process continues until the network stabilizes on all the input patterns. The stable state so achieved is described as *resonance*, a concept used in the *Adaptive Resonance Theory*.

Figure for BAM Network

You will find in Figure 8.1 in Chapter 8, the layout of the nodes that represent the neurons in a Bidirectional Associative Memory network and the connections between them. There are two fields of neurons. The network is fully connected with feedback connections and forward connections. There are no lateral or recurrent connections.

Brain-State-in-a-Box

Introduced by James Anderson and others, this network differs from the single- layer fully connected Hopfield network in that Brain-State-in-a-Box uses recurrent connections as well. Each neuron has a connection to itself. With target patterns available, a modified *Hebbian learning* rule is used.

The adjustment to a connection weight is proportional to the product of the desired output and the error in the computed output. This network is adept at noise tolerance, and it can accomplish pattern completion. Figure 5.7 shows a Brain-State-in-a-Box network.

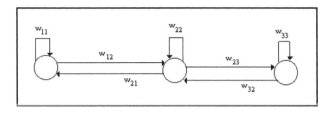

FIGURE 5.7   *Brain-State-in-a-Box.*

Neocognitron

Compared to all the other neural network models, Fukushima's Neocognitron is more complex, sophisticated, and ambitious. It demonstrates the advantages of a multilayered network. The Neocognitron is one of the best models for recognizing handwritten symbols. Many pairs of layers called the *S layer*, for *simple layer*, and *C layer*, for *complex layer*, are used. Within each S layer are several planes containing simple cells. Similarly, there are within each C layer, an equal number of planes containing complex cells. The input layer does not have this arrangement and is like an input layer in any other neural network.

The number of planes of simple cells and of complex cells within a pair of S and C layers being the same, these planes are paired, and the complex plane cells process the outputs of the simple plane cells. The simple cells are trained so that the response of a simple cell corresponds to a specific portion of the input image. If the same part of the image occurs with some distortion, in terms of scaling or rotation, a different set of simple cells respond to it. The complex cells output to indicate that some simple cell they correspond to did fire. While simple cells respond to what is in a contiguous region in the image, complex cells respond on the basis of a larger region. As the process continues to the output layer, the C-layer component of the output layer responds, corresponding to the entire image presented in the beginning at the input layer.

Adaptive Resonance Theory

ART1 is the first model for adaptive resonance theory for neural networks developed by Gail Carpenter and Stephen Grossberg. This theory was developed to address the stability–plasticity dilemma. The network is supposed to be plastic enough to learn an important pattern. But at the same time it should remain stable when, in short-term memory, it encounters some distorted versions of the same pattern.

ART1 model has A and B field neurons, a gain, and a reset as shown in Figure 5.8. There are top-down and bottom-up connections between neurons of fields A and B. The neurons in field B have lateral connections as well as recurrent connections. That is, every neuron in this field is connected to every other neuron in this field, including itself, in addition to the connections to the neurons in field A. The external input (or bottom-up signal), the top-down signal and the gain constitute three elements of a set, of which at least two should be a +1 for the neuron in the A field to fire. This is what is termed as the *two-thirds rule*. Initially, therefore, the gain would be set to +1. The idea of a single winner is also employed in the B field. The gain would not contribute in the top-down phase; actually, it will inhibit. The two-thirds rule helps move toward stability once resonance is obtained. A vigilance parameter ρ is used to determine the parameter reset. Vigilance parameter corresponds to what degree the resonating category can be predicted. The part of the system that contains gain is called the *attentional subsystem*, whereas the rest, the part that contains reset, is termed the *orienting subsystem*. The top-down activity corresponds to the orienting subsystem , and the bottom-up activity relates to the attentional subsystem.

In ART1 classification of an input pattern in relation to stored patterns is attempted, and if unsuccessful, a new stored classification is generated. Training is unsupervised. There are two versions of training: slow and fast. They differ in the extent to which the weights are given the time to reach their eventual values. Slow training is governed by differential equations, and fast training by algebraic equations. ART2 is the analog counterpart of ART1, which is for discrete cases. These are self-organizing neural networks, as you can surmise from the fact that training is present but unsupervised.

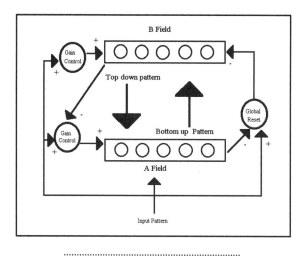

FIGURE 5.8   *The ART1 network.*

Summary

The basic concepts of neural network layers, connections, weights, inputs, and outputs have been discussed. An example of how adding another layer of neurons in a network can solve a problem that could not be solved without it is given in detail. A number of neural network models are introduced briefly. Learning and training, which form the determinants of a neural network behavior has not been included here, but will be discussed in the following chapter.

Chapter
6

Learning, Self-Organization, and Resonance

Learning and Self-Organization

nsupervised learning and self-organization are closely related. Unsupervised learning was mentioned in Chapter 1, along with supervised learning. Learning comes from a training process, which is either supervised or unsupervised. Training in supervised learning takes the form of external exemplars being provided. The network has to compute the correct weights for the connections for neurons in some layer or the other. Self-organization implies unsupervised learning. It was described as a characteristic of a neural network model, ART1, based on adaptive resonance theory. With the **winner take all**

93

criterion each neuron of field B learns a distinct classification. The winning neuron in a layer, in this case the field B, is the one with the largest activation, and it is the only neuron in that layer that is allowed to fire. Hence, the name winner take all.

Learning

A network can learn when training is used, or the network can learn also in the absence of training. The difference between supervised and unsupervised training is that, in the former case, external prototypes are used as target outputs for specific inputs and the network is given a learning algorithm to follow and calculate the correct connection weights. Unsupervised learning is the sort of learning that takes place without a teacher. For example, when you are finding your way out of a labyrinth, no teacher is present. You learn from the responses or events that develop as you try to feel your way through the maze. For neural networks, in the unsupervised case, a learning algorithm may be given but target outputs are not given.

When a neural network model is developed and an appropriate learning algorithm is proposed, it would be based on the theory supporting the model. Since the dynamics of the operation of the neural network is under study, the learning equations are initially formulated in terms of differential equations. After solving the differential equations, and using any initial conditions that are available, the algorithm is simplified to consist of an algebraic equation for the changes in the weights. These simple forms of learning equations are available for your neural networks. At this point of our discussion you need to know what learning algorithms are available, and what they look like. Basically, there are two rules for learning—*Hebbian learning* and the *delta rule*. Adaptations of these by simple modifications to suit a particular context generate some other learning rules.

Hebb's Rule

Learning algorithms are usually referred to as *learning rules*. The foremost such rule is due to Donald Hebb. Hebb's rule is a statement about how the firing of one neuron, which has a role in the determination of the activation of another neuron, affects the first neuron's influence on the activation of the second neuron, especially if it is done in a repetitive manner. As a learning rule, Hebb's observation translates into a formula for the difference in a connection weight between two neurons from one iteration to the next, as a constant μ times the product of activations of the two neurons. How a connection weight is to be modified is what the learning rule suggests. In the case of Hebb's rule, it is adding the quantity $\mu a_i a_j$, where a_i is the activation of the ith neuron, and a_j is the activation of the jth neuron to the connection weight between the ith and jth neurons. The constant μ itself is referred to as the *learning rate*. The following equation using the notation just described, states it succinctly.

$$\Delta w_{ij} = \mu a_i a_j$$

As you can see, the learning rule derived from Hebb's rule is quite simple and is used in both simple and more involved networks. Some modify this rule by replacing the quantity a_i with its deviation from the average of all a's and, similarly, replacing a_j by a corresponding quantity. Such rule variations can yield rules better suited to different situations.

For example, the output of a neural network being the activations of its output layer neurons, the Hebbian learning rule in the case of a perceptron takes the form of adjusting the weights by adding μ times the difference between the output and the target. Sometimes a situation arises where some unlearning is required for some neurons. In this case a reverse Hebbian rule is used in which the quantity $\mu a_i a_j$ is subtracted from the connection weight under question, which in effect is employing a negative learning rate.

In the Hopfield network of Chapter 1, there is a single layer with all neurons fully interconnected. Suppose each neuron's output is either a + 1 or a – 1. If we take μ = 1 in the Hebbian rule, the resulting modification of the connection weights can be described as follows: add 1 to the weight, if both neuron outputs match, that is, both are +1 or –1. And if they do not match (meaning one of them has output +1 and the other has –1), then subtract 1 from the weight.

Delta Rule

The delta rule is also known as the *least mean squared error rule* (LMS). You take the square of the errors between the target or desired values and computed values, and take the average to get the mean squared error. This quantity is to be minimized. For this, you first realize that it is a function of the weights themselves, since the computation of output uses them. The set of values of weights that minimizes the mean squared error is what is needed for the next cycle of operation of the neural network. Having worked this out mathematically, and having compared the weights thus found with the weights actually used, one determines their difference and gives it in the delta rule. So the delta rule, which is also the rule used first by Widrow and Hoff, in the context of learning in neural networks, is stated as an equation defining the change in the weights to be affected. Thus, if in the cycle of operation corresponding to time t + 1, the weight on the connection between the ith neuron in one layer and the jth neuron it is connected to is $w_{ij}(t + 1)$, and if $w_{ij}(t)$ is the corresponding value at t, then their difference denoted by Δw_{ij} is given by:

```
Δwᵢⱼ = 2µxᵢ(desired output value - computed output value)ⱼ
```

Here, μ is the learning rate, which is positive and much smaller than 1, and x_i is the ith component of the input vector.

Other Learning Rules

Generalized Delta Rule

This moniker belongs to the Backpropagation algorithm. While the delta rule uses local information on error, the *generalized delta rule* uses error information that is not local. It is designed to minimize the total of the squared errors of the output neurons. In trying to achieve this minimum, the *steepest descent* method, which uses the gradient of the weight surface, is used. The weights of the output layer neurons are adjusted differently from the way the weights for the hidden layer neurons are modified.

Self-organization

Self-organization is also unsupervised learning. It is self-adaptation of a neural network. Without target outputs, the best possible response to a given input signal is to be generated. The connection weights are modified through different iterations of network operation, and the network capable of self-organizing creates on its own the best possible set of outputs for the given inputs. This happens in the models in which, a layer of neurons on the lines of a model by Kohonen is used.

Kohonen's *Linear Vector Quantizer* (LVQ) described briefly below is later extended as a self-organizing feature map. Self-organization is also learning, but without supervision; it is a case of self-training. Kohonen's Topology Preserving Maps illustrate self-organization by a neural network. In these cases, certain subsets of output neurons respond to certain subareas of the inputs, so that the firing within one subset of neurons indicates the presence of the corresponding subarea of the input. This is a useful paradigm in applications such as speech recognition. Winner take all strategy used in ART1 also facilitates self-organization.

Learning Vector Quantizer

The *Learning Vector Quantizer* (LVQ) of Kohonen is a self-organizing network. It classifies input vectors on the basis of a set of stored or reference vectors. The B field neurons are also called *grandmother cells*. Each of them represents a specific class in the reference vector set. Either supervised or unsupervised learning can be imposed. Figure 6.1 shows the layout for LVQ.

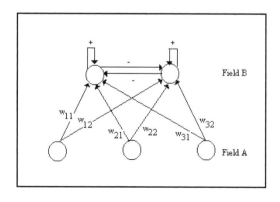

FIGURE 6.1 *Layout for Learning Vector Quantizer.*

Resonance

ART1 is the first neural network model based on adaptive resonance theory of Carpenter and Grossberg. When you have a pair of patterns such that when one of them is input to a neural network , the output turns out to be the other pattern in the pair, and if this happens consistently in both directions, you may describe it as resonance. We discuss in Chapter 8 bidirectional associative memories and resonance. By the time training is completed, and learning is through, many other pattern pairs would have been presented to the network as well. If changes in the short-term memory do not disturb or affect the long-term memory, the network shows adaptive resonance. The ART1 model is designed to maintain it. You realize of course that these questions relate largely to stability.

Stability

Learning, convergence, and stability are matters of much interest in that sequential order. As learning is taking place, you want to know if the process is going to halt at some appropriate point, which is a question of convergence. Is what is learned stable, or will the network have to learn all over again, as each new event occurs? These questions have their answers within a mathematical model with differential equations, developed to describe a learning algorithm. Proofs showing stability are part of the model inventor's task. One particular tool that aids in the process of showing convergence is the idea of state energy, or cost, to describe whether the direction the process is taking can lead to convergence.

The *Lyapunov* function, discussed later in this chapter, is found to provide the right energy function, which can be minimized during the operation of the neural network. This type of function has the property that the value gets smaller with every change in the state of the system, thus assuring that a minimum will be reached eventually. The Lyapunov function is discussed further because of its significant utility for neural network models, but briefly because the level of mathematics involved is high. Fortunately, simple forms are derived and put into learning algorithms for neural networks. The high level mathematics is utilized in making the proofs to show the viability of the models.

Alternatively, temperature relationships can be used, as in the case of the Boltzman Machine, or any other well-suited cost function such as a function of distances used in the formulation of the *Traveling Salesman Problem*, in which the total distance for the tour of the traveling salesman is to be minimized, can be employed. The Traveling Salesman Problem is an important and well-known problem. A set of cities are to be visited by the salesman, each only once, and the aim is to devise a tour that minimizes the total distance traveled. The search continues for an efficient algorithm for this problem. Some algorithms solve the problem in a large number of situations, but not in every situation. The neural network approach can also work for the Traveling Salesman Problem.

Convergence

Suppose you have a criterion such as energy to be minimized or cost to be decreased, and you know the optimum level for this criterion. If the network achieves the optimum value in a finite number of steps, you have convergence for the operation of the network. Or, if you are making pairwise associations of patterns, there is the prospect of convergence if after each cycle of the network operation, the number of errors is decreasing.

It is also possible that convergence is slow, so much so that it may seem to take forever to achieve the convergence state. In that case, you should specify a tolerance value and require that the criterion be achieved within that tolerance, avoiding a lot of computing time. You may also introduce a *momentum* parameter to further change the weight and thereby speed up the convergence. One technique used is to add a portion of the previous change in weight.

Instead of converging, the operation may result in oscillations. The weight structure may keep changing back and forth. Learning will never cease. Learning algorithms need to be analyzed in terms of convergence as being an essential property of an algorithm.

Lyapunov Function

Neural networks being dynamic systems in the learning and training phase of their operation, and convergence being an essential feature, it was necessary for the researchers developing the models and their learning algorithms to find a provable criterion for convergence in a dynamic system. The Lyapunov function, mentioned previously, turned out to be the most convenient and appropriate function. It is also referred to as the energy function. The function decreases as the system states change. Such a function needs to be found and watched as the network operation continues from one cycle to another. Usually it involves a quadratic form. The least mean squared error is an example of such a function. Lyapunov function usage assures a stability of the system that cannot occur without convergence. It is convenient to have one value, that of the Lyapunov function specifying the system behavior. For example, in the Hopfield network, the energy function is a constant times the sum of products of outputs of different neurons and the connection weight between them. Since pairs of neuron outputs are multiplied in each term, the entire expression is a quadratic form.

Boltzman/Cauchy Machines

The Boltzman machine (and Cauchy machine) uses probabilities and statistical theory, along with an energy function representing temperature. The learning is probabilistic and is called *simulated annealing*. At different temperature levels, a different number of iterations in processing are used, and this constitutes an annealing schedule. Use of probability distributions is for the goal of reaching a state of global minimum of energy. Boltzman distribution and Cauchy distribution are probability distributions used in this process. It is obviously desirable to reach a global minimum, rather than settling down at a local minimum, as it can happen while using a Lyapunov function.

Figure 6.2 clarifies the distinction between a local minimum and a global minimum. In this figure you find the graph of an energy function and two points A and B. These points show that the energy levels there are smaller than the energy levels at any point in their vicinity, so you can say they represent points of minimum energy. The overall or global minimum, as you can see, is at point B, where the energy level is smaller than that even at point A, so A corresponds only to a local minimum. It is desirable to get to B and not get stopped at A itself, in the pursuit of a minimum for the energy function. If point C is reached, one would like the further movement to be toward B and not A. Similarly, if a point near A is reached, the subsequent movement should avoid reaching or settling at A but proceed to B. Perturbation techniques are useful for these considerations.

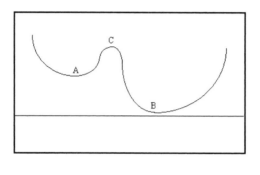

FIGURE 6.2   *Local and Global minima.*

Clamping Probabilities

Sometimes in simulated annealing, first a subset of the neurons in the network are associated with some inputs, and another subset of neurons are associated with some outputs, and these are clamped with probabilities, which are not changed in the learning process. Then the rest of the network is subjected to adjustments. Updating is not done for the clamped units in the network. This training procedure of Geoffrey Hinton and Terrence Sejnowski provides an extension of the Boltzman technique to more general networks.

Other Needs for a Neural Network Model

Besides the applications for which a Neural Network is intended, and depending on these applications, you need to know certain aspects of the model. The length of encoding time and the length of learning time are among the important considerations. These times could be long but should not be prohibitive. Whether or not generalization is feasible is another consideration. How the network behaves with new inputs is something to know about. Some may need to be trained all over again. Some tolerance for distortion in input patterns is desirable, where relevant. Restrictions on the format of inputs, or to their linearity is to be known. One of the advantages of neural networks is that they can deal with nonlinear functions better than traditional algorithms can. Ability to store a number of patterns, or needing more and more neurons in the output field with increasing number of input patterns are the kind of aspects addressing the capabilities of a network and also its limitations. This kind of information is available in the existing literature.

Summary

This chapter dealt with the following concepts:

- Learning
- Self-organization
- Resonance
- Convergence
- Stability

Stability on the one hand can relate to the concept of convergence, and on the other hand, that of stable recall.

Chapter 7

Backpropagation

Backpropagation Network

he backpropagation network is a very popular model in neural networks. It does not have feedback connections, but errors are backpropagated during training. Least mean squared error is used. Many applications can be formulated for using a backpropagation network, and the methodology has been a model for most multilayer neural networks. Errors in the output determine measures of hidden layer output errors, which are used as a basis for adjustment of connection weights between the input and hidden layers. Adjusting the two sets of weights between the pairs of layers and recalculating the outputs is an iterative process that is carried on until the errors fall below a tolerance level. Learning rate parameters scale the adjustments to weights. A momentum parameter can also be used in scaling the adjustments from a previous iteration and adding to the adjustments in the current iteration.

103

Mapping

The backpropagation network maps the input vectors. Pairs of input and output vectors are chosen to train the network first. Once training is completed, the weights are set and the network can be used to find outputs for new inputs. The number of neurons in the input layer determines the dimension of the inputs, and the number of neurons in the output layer determines the dimension of the outputs. If there are k neurons in the input layer and m neurons in the output layer, then this network can make a mapping of the k-dimensional space to an m-dimensional space. Of course, what that mapping is depends on what pair of patterns or vectors are used as exemplars to train the network. Once trained, the network would give you the image of a new input vector under this mapping. Knowing what mapping you want the backpropagation network to be trained for tells you the dimensions of the input space and the output space, so that you can determine the numbers of neurons to have in the input and output layers.

Layout

The architecture of a backpropagation network is shown in Figure 7.1. While there can be many hidden layers, we will illustrate this network with only one hidden layer. Also, the number of neurons in the input layer and that in the output layer are determined by the dimensions of the input and output patterns, respectively. As for the hidden layer, it is not easy to determine how many neurons are needed. In order to avoid cluttering the figure, we will show the layout in Figure 7.1 with five input neurons, three neurons in the hidden layer, and four output neurons with a few representative connections.

Therefore we have three fields, one for input neurons, one for hidden processing elements, and one for the output neurons. As already stated, connections are for feed forward activity. There are connections from every neuron in fields A to every one in field B, and, in turn, from every neuron in field B to every neuron in field C. Thus there are two sets of weights, those figuring in the activations of hidden layer neurons, and those that help determine the output neuron activations. In training, all of these weights are adjusted by considering what can be called a *cost function* in terms of the error in the computed output pattern and the desired output pattern.

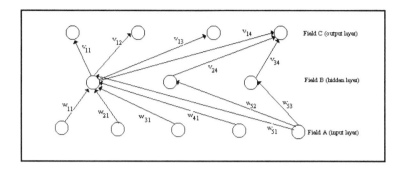

Field C (output layer)

Field B (hidden layer)

Field A (input layer)

FIGURE 7.1 *Layout of a Backpropagation Network.*

Training

The backpropagation network undergoes supervised training, with a finite number of pattern pairs consisting of an input pattern and a desired or target output pattern. An input pattern is presented at the input layer. The neurons here pass the pattern digits to the next layer neurons, which are in a hidden layer. The outputs of the hidden layer neurons are obtained by using perhaps a bias, and also a threshold function with the activations determined by the weights and the inputs. These hidden layer outputs become inputs to the output neurons, which also process using possibly a **bias** and a **threshold** function with their activations to determine the final output from the network.

The computed pattern and the input pattern are compared, a function of this error for each component of the pattern is determined, and adjustment to weights of connections between the hidden layer and the output layer is computed. A similar computation, still based on the error in the output, is made for the connection weights between the input and hidden layers. The process is then repeated as many times as needed until the error is within a prescribed tolerance. This procedure is repeated with each pattern pair assigned for training the network.

There can be one or more than one learning rate parameters used in training in a backpropagation network. You can use one with each set of weights between consecutive layers.

NOTE

Illustration: Adjustment of Weights of Connections from a Neuron in the Hidden Layer

We will be as specific as is needed to make the computations clear. First recall that the activation of a neuron in a layer other than the input layer, is the sum of products of its inputs and the weights corresponding to the connections that bring in those inputs. Let us discuss the jth neuron in the hidden layer. Let us be specific and say j = 2. If the input pattern is (1.1, 2.4, 3.2, 5.1, 3.9) and the target output pattern is (0.52, 0.25, 0.75, 0.97). Let the weights be given for the second hidden layer neuron by the vector (–0.33, 0.07, –0.45, 0.13, 0.37). The activation will be the quantity,

```
(-0.33 * 1.1) + (0.07 * 2.4) + (-0.45 * 3.2) + (0.13 * 5.1) +
(0.37 * 3.9) = 0.471
```

Now add to this an optional bias, or threshold value of, say, 0.679, to give 1.15. If we use the sigmoid function given by

```
1 / ( 1+ exp(-x) ),
```

with x = 1.15, we get the output of this hidden layer neuron as 0.7595.

We are taking values to a few decimal places only, unlike the precision that can be obtained on a computer.

N O T E

We need the computed output pattern also. Let us say, it turns out to be (0.61, 0.41, 0.57, 0.53), while the desired pattern is (0.52, 0.25, 0.75, 0.97). Obviously, there is a discrepancy between what is desired and what is computed. The component-wise differences are given in the vector (–0.09, –0.16, 0.18, 0.44). We use these to form another vector where each component is a product of the error component, corresponding computed pattern component, and the complement of the latter with respect to 1. For example, for the first component, error is –0.09, computed pattern component is 0.61, and its complement is 0.39. Multiplying these together, we get –0.02. Calculating the other components similarly, we get the vector (–0.02, –0.04, 0.04, 0.11).

We need now the weights on the connections between the second neuron in the hidden layer that we are concentrating on, and the different output neurons. Let us say these weights are given by the vector (0.85, 0.62, –0.10, 0.21). The error of the second neuron in the hidden layer is now calculated as below, using its output.

```
error = 0.7595 * (1 - 0.7595) * ( (0.85 * -0.02) + (0.62 *
        -0.04) + ( -0.10 * 0.04) + (0.21 * 0.11)) = -0.0041.
```

Next, we need the learning rate parameter for this layer. Let us set it as 0.2. We multiply this by the output of the second neuron in the hidden layer, to get 0.1519. Each of the components of the vector (–0.02, –0.04, 0.04, 0.11) is multiplied now by 0.1519, which our latest computation gave. The result is a vector that gives the adjustments to the weights on the connections that go from the second neuron in the hidden layer to the output neurons. These values are given in the vector (–0.003, –0.006, 0.006, 0.017). After these adjustments are added, the weights to be used in the next cycle on the connections between the second neuron in the hidden layer and the output neurons become those in the vector (0.847, 0.614, –0.094, 0.227).

Illustration: Adjustment of Weights of Connections from a Neuron in the Input Layer

Let us look at how adjustments are calculated for the weights on connections going from the ith neuron in the input layer to neurons in the hidden layer. Let us take specifically i = 3, for illustration.

Much of the information we need is already obtained in the previous discussion for the second hidden layer neuron. We have the errors in the computed output as the vector (–0.09, –0.16, 0.18, 0.44), and we obtained the error for the second neuron in the hidden layer as –0.0041, which was not used above. Just as the error in the output is propagated back to the neurons in the hidden layer, error computed for the hidden neuron outputs, which is obtained as a sort of by-product, can be propagated to the input layer neurons.

To determine the adjustments for the weights on connections between the input and hidden layers, we need the errors determined for the outputs of hidden layer neurons, a learning rate parameter, and the activations of the input neurons, which are just the inputs. Let us take the learning rate parameter to be 0.15. Then the weight adjustments for the connections from the third input neuron to the hidden layer neurons are obtained by multiplying the particular hidden layer neuron's output error by the learning rate parameter and by the input component from the input neuron. The adjustment for the weight on the connection from the third input neuron to the second hidden layer neuron is 0.15 * 3.2 * –0.0041, which works out to –0.002.

If the weight on this connection is, say, –0.45, then adding the adjustment of –0.002, we get the modified weight of –0.452, to be used in the next iteration of the network operation. Similar calculations are made to modify all other weights as well.

Adjustments to Threshold Values or Biases

The bias or the threshold value we added to the activation, before applying the threshold function to get the output of a neuron, will also be adjusted based on the error being propagated back. The needed values for this are in the previous discussion.

The adjustment for the threshold value of a neuron in the output layer is obtained by multiplying the calculated error (not just the difference) in the output at the output neuron and the learning rate parameter used in the adjustment calculation for weights at this layer. In our previous example, we have the learning rate parameter as 0.2, and the error vector as (–0.02, –0.04, 0.04, 0.11), so the adjustments to the threshold values of the four output neurons are given by the vector (–0.004, –0.008, 0.008, 0.022). These adjustments are added to the current levels of threshold values at the output neurons.

The adjustment to the threshold value of a neuron in the hidden layer is obtained similarly by multiplying the learning rate with the computed error in the output of the hidden layer neuron. Therefore for the second neuron in the hidden layer, the adjustment to its threshold value is calculated as 0.15 * –0.0041, which is –0.0006. Add this to the current threshold value of 0.679 to get 0.6784, which is to be used for this neuron in the next cycle of operation of the neural network.

Notation and Equations

You have just seen an example of the process of training in the backpropagation network, described in relation to one hidden layer neuron and one input neuron. There were a few vectors that were shown and used, but perhaps not made easily identifiable. We therefore introduce some notation and describe the equations that were implicitly used in the example.

Notation

Let us talk about two matrices whose elements are the weights on connections. One matrix refers to the interface between the input and hidden layers, and the second refers to that between the hidden layer and the output layer. Since connections exist from each neuron in one layer to every neuron in the next layer, there is a vector of weights on the connections going out from any one neuron. Putting this vector into a row of the matrix, we get as many rows as there are neurons from which connections are established.

Let M_1 and M_2 be these matrices of weights. Then what does $M_1[i][j]$ represent? It is the weight on the connection between the ith input neuron to the jth neuron in the hidden layer. Similarly, $M_2[i][j]$ denotes the weight on the connection between the ith neuron in the hidden layer and the jth output neuron.

Next, we will use x, y, z for the outputs of neurons in the input layer, hidden layer, and output layer, respectively, with a subscript attached to denote which neuron in a given layer we are referring to. Let P denote the desired output pattern, with p_i as the components. Let m be the number of input neurons, so that according to our notation, (x1, x2, ..., xm) will denote the input pattern. If P has, say, r components, the output layer needs r neurons. Let the number of hidden layer neurons be n. Let λ be the learning rate parameter for the hidden layer, and μ, that for the output layer. Let θ with the appropriate subscript represent the threshold value or bias for a hidden layer neuron, and τ with an appropriate subscript refer to the threshold value of an output neuron.

Let the errors in output at the output layer be denoted by e_j's and those at the hidden layer by t_i's. If we use as a Δ prefix of any parameter, then we are looking at the change in or adjustment to that parameter. Also, the **thresholding** function we would use is the **sigmoid** function, $f(x) = 1 / (1 + \exp(-x))$.

Equations

output of jth hidden layer neuron:

$$y_j = f((\Sigma_i x_i M_1[i][j]) + \theta j) \qquad (7.1)$$

output of jth output layer neuron:

$$z_j = f((\Sigma_i y_i M_2[i][j]) + \tau_j) \qquad (7.2)$$

ith component of vector of output differences: desired value-computed value $= P_i - z_i$

ith component of output error at the output layer:

$$e_i = z_i (1 - z_i) (P_i - z_i) \qquad (7.3)$$

ith component of output error at the hidden layer:

$$t_i = y_i (1 - y_i) (\Sigma j M_2[i][j] e_j) \qquad (7.4)$$

adjustment for weight between ith neuron in hidden layer and jth output neuron:

$$\Delta M_2[i][j] = \mu y_i e_j \qquad (7.5)$$

adjustment for weight between ith input neuron and jth neuron in hidden layer:

$$M_1[i][j] = \lambda x_i t_j \qquad (7.6)$$

adjustment to the threshold value or bias for the jth output neuron:

$$\Delta \tau_j = \mu e_j$$

adjustment to the threshold value or bias for the jth hidden layer neuron:

$$\Delta \theta_j = \lambda e_j$$

use of momentum parameter γ: Instead of equations 7.5 and 7.6, use

$$\Delta M_2[\ i\ ][\ j\ ]\ (\ t\ )\ =\ \mu y_i e_j\ +\ \gamma \Delta M_2[\ i\ ][\ j\ ]\ (\ t\ -\ 1\ ) \qquad (7.7)$$

and

$$\Delta M_1[\ i\ ][\ j\ ]\ (\ t\ )\ =\ \lambda x_i t_j\ +\ \gamma \Delta M_1[\ i\ ][\ j\ ]\ (t\ -\ 1) \qquad (7.8)$$

C++ Implementation of a Backpropagation Simulator

The backpropagation simulator of this chapter has the following design objectives:

1. Allow the user to specify the number and size of all layers.
2. Allow the use of one or more hidden layers.
3. Be able to save and restore the state of the network.
4. Run from an arbitrarily large training data set or test data set.
5. Query the user for key network and simulation parameters.
6. Display key information at the end of the simulation.
7. Demonstrate the use of some C++ features.

A Brief Tour of How to Use the Simulator

In order to understand the C++ code, let us have an overview of the functioning of the program.

There are two modes of operation in the simulator. The user is queried first for which mode of operation is desired. The modes are *Training* mode and *Nontraining* mode (*Test* mode).

Training Mode

Here, the user provides a training file in the current directory called *training.dat*. This file contains exemplar pairs, or patterns. Each pattern has a set of inputs followed by a set of outputs. Each value is separated by one or more spaces. As a convention, you can use a few extra spaces to separate the inputs from the outputs. Here is an example of a training.dat file that contains two patterns:

```
0.4 0.5 0.89      -0.4 -0.8
0.23 0.8 -0.3      0.6 0.34
```

In this example, the first pattern has inputs 0.4, 0.5, and 0.89, with an expected output of –0.4 and –0.8. The second pattern has inputs of 0.23, 0.8 and –0.3 and outputs of 0.6 and 0.34. Since there are three inputs and two outputs, the input layer size for the network must be three neurons and the output layer size must be two neurons. Another file that is used in training is the weights file. Once the simulator reaches the **error tolerance** that was specified by the user, or the **maximum number of iterations**, the simulator saves the state of the network, by saving all of its weights in a file called *weights.dat*. This file can then be used subsequently in another run of the simulator in **Nontraining** mode. To provide some idea of how the network has done, information about the total and average error is presented at the end of the simulation. In addition, the output generated by the network for the last pattern vector is provided in an output file called *output.dat*.

Nontraining Mode (Test Mode)

In this mode, the user provides test data to the simulator in a file called *test.dat*. This file contains only input patterns. When this file is applied to an already trained network, an output.dat file is generated, which contains the outputs from the network for all of the input patterns. The network goes through one cycle of operation in this mode, covering all the patterns in the test data file. To start up the network, the weights file, weights.dat is read to initialize the state of the network. The user must provide the same network size parameters used to train the network.

Operation

The first thing to do with your simulator is to train a network with an architecture you choose. You can select the number of layers and the number of hidden layers for your network. Keep in mind that the input and output layer sizes are dictated by the input patterns you are presenting to the network and the outputs you seek from the network. Once you decide on an architecture, perhaps a simple three-layer network with one hidden layer, you prepare training data for it and save the data in the training.dat file. After this you are ready to train. You provide the simulator with the following information:

◆ The mode you are using (select 1 for training).

◆ The values for the error tolerance and the learning rate parameter, lambda or beta.

◆ The maximum number of cycles, or passes through the training data you'd like to try.

◆ The number of layers (between three and five, three implies one hidden layer whereas five implies three hidden layers).

◆ The size for each layer, from the input to the output.

The simulator then begins training and reports the current cycle number and the average error for each cycle. *You should watch the error to see that it is decreasing with time.* If it is not, you should restart the simulation, because this will start with a brand new set of random weights and give you another, possibly better, solution. Once the simulation is done you will see information about the number of cycles and patterns used, and the total and average error that resulted. The weights are saved in the weights.dat file. You can rename this file to use this particular state of the network later. You can infer the size and number of layers from the information in this file, as will be shown in the next section for the weights.dat file format. You can have a peek at the output.dat file to see the kind of training result you have achieved. To get a full-blown accounting of each pattern and the match to that pattern, copy the training file to the test file and delete the output information from it. You can then run **Test** mode to get a full list of all the input stimuli and responses in the output.dat file.

Summary of Files Used in the Backpropagation Simulator

Here are the files again and what they are used for:

◆ **weights.dat** You can look at this file to see the weights for the network. It shows the layer number followed by the weights that feed into the layer. The first layer, or input layer, layer zero, does not have any weights associated with it. An example of the weights.dat file is shown as follows for a network with three layers of sizes 3, 5, and 2. Note that the row width for layer n matches the column length for layer n + 1:

```
1 -0.199660-0.859660-0.339660-0.2596600.520340
1 0.292860-0.4871400.212860-0.967140-0.427140
1 0.542106-0.1778940.322106-0.9778940.562106
2 -0.175350-0.835350
2 -0.330167-0.250167
2 0.5033170.283317
2 -0.4771580.222842
2 -0.928322-0.388322
```

In this weights file the row width for layer 1 is 5, corresponding to the output of that (middle) layer. The input for the layer is the column length, which is 3, just as specified. For layer 2, the output size is the row width, which is 2, and the input size is the column length, 5, which is the same as the output for the middle layer. You can read the weights file to find out how things look.

◆ **training.dat** This file contains the input patterns for training. You can have as large a file as you'd like without degrading the performance of the simulator. The simulator caches data in memory for processing. This is to improve the speed of the simulation since disk accesses are expensive in time. A data buffer, which has a maximum size specified in a **#define** statement in the program, is filled with data from the training.dat file whenever data is needed. The format for the training.dat file has been shown in the **Training** mode section.

◆ **test.dat** The test.dat file is just like the training.dat file but without expected outputs. You use this file with a trained neural network in **Test** mode to see what responses you get for untrained data.

◆ **output.dat** The output.dat file contains the results of the simulation. In **Test** mode, the input and output vectors are shown for all pattern vectors. In the **Simulator** mode, the expected output is also shown, but only the last vector in the training set is presented, since the **Training** set is usually quite large.

Shown here is an example of an output file in **Training** mode:

```
for input vector:
0.400000  -0.400000
output vector is:
0.880095
expected output vector is:
0.900000
```

C++ Classes and Class Hierarchy

So far, you can see how we address most of the objectives outlined for this program. The only objective left is regarding the demonstration of some C++ features. In this program we use a class hierarchy with the **inheritance** feature. Also, we use polymorphism with *dynamic* binding and function overloading with *static* binding.

First let us look at the class hierarchy used for this program. Figure 7.2 shows this hierarchy. An *abstract* class is a class that is never meant to be instantiated as an object, but serves as a base class from which others can inherit functionality and interface definitions. The layer class is such a class. You will see shortly that one of its functions is set = zero, which indicates that this class is an abstract base class. From the layer class are two branches. One is the **input_layer** class, and the other is the **output_layer** class. The middle layer class is very much like the output layer in function and so inherits from the **output_layer** class.

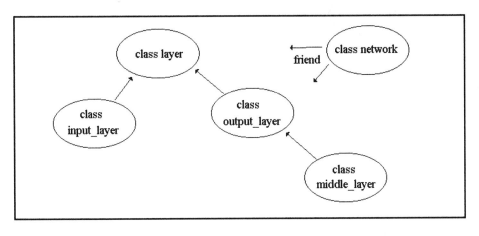

FIGURE 7.2 *Class hierarchy used in the backpropagation simulator*

Function overloading can be seen in the definition of the **calc_error()** function. It is used in the **input_layer** with no parameters, while it is used in the **output_layer** (which the **input_layer** inherits from) with one parameter. Using the same function name is not a problem, and this is referred to as *overloading*. Besides function overloading, you may also have operator overloading, which is using an operator that performs some familiar function like + for addition, for another function, say, vector addition.

When you have overloading with the same parameters and the keyword *virtual*, then you have the potential for *dynamic binding*, which means that you determine which overloaded function to execute at run time and not at compile time, or *static binding*. If you put a bunch of C++ objects in an array of pointers to the **base** class, and then go through a loop that indexes each pointer and executes an overloaded virtual function that that pointer is pointing to, then you will be using dynamic binding. This is exactly the case in the function called **calc_out()**, which is declared with the virtual keyword in the **layer base** class. Each descendant of layer can provide a version of **calc_out()**, which differs in functionality from the base class, and the correct function will be selected at run time based on the object's identity. In this case **calc_out()**, which is a function to calculate the outputs for each layer, is different for the input layer than for the other two types of layers.

Let's look at some details in the header file in Listing 7.1:

Listing 7.1   Header file for the backpropagation simulator

```
// layer.h    V.Rao, H. Rao
// header file for the layer class hierarchy and
// the network class

#define MAX_LAYERS   5
#define MAX_VECTORS 100


class network;

class layer
{

protected:

    int num_inputs;
    int num_outputs;
    float *outputs;  // pointer to array of outputs
    float *inputs;   // pointer to array of inputs, which
          // are outputs of some other layer
```

```
        friend network;

    public:

        virtual void calc_out()=0;
    };

    class input_layer: public layer
    {

    private:

    public:

        input_layer(int, int);
        ~input_layer();
        virtual void calc_out();

    };

    class middle_layer;

    class output_layer: public layer
    {
    protected:

        float * weights;
        float * output_errors; // array of errors at output
        float * back_errors; // array of errors back-propagated
        float * expected_values;   // to inputs
        friend network;

    public:

        output_layer(int, int);
        ~output_layer();
        virtual void calc_out();
        void calc_error(float &);
        void randomize_weights();
        void update_weights(const float);
        void list_weights();
        void write_weights(int, FILE *);
        void read_weights(int, FILE *);
        void list_errors();
        void list_outputs();
    };

    class middle_layer: public output_layer
    {
```

```
private:

public:
    middle_layer(int, int);
    ~middle_layer();
    void calc_error();
};


class network

{

private:

    layer *layer_ptr[MAX_LAYERS];
    int number_of_layers;
    int layer_size[MAX_LAYERS];
    float *buffer;
    fpos_t position;
    unsigned training;

public:
    network();
    ~network();
    void set_training(const unsigned &);
    unsigned get_training_value();
    void get_layer_info();
    void set_up_network();
    void randomize_weights();
    void update_weights(const float);
    void write_weights(FILE *);
    void read_weights(FILE *);
    void list_weights();
    void write_outputs(FILE *);
    void list_outputs();
    void list_errors();
    void forward_prop();
    void backward_prop(float &);
    int fill_IObuffer(FILE *);
    void set_up_pattern(int);

};
```

Details of the Backpropagation Header File

At the top of the file, there are two **#define** statements, which are used to set the maximum number of layers that can be used, currently five, and the maximum number of training or test vectors that can be read into an I/O buffer. This is currently 100. You can increase the size of the buffer for better speed at the cost of increased memory usage.

The following are definitions in the **layer base** class. Note that the number of inputs and outputs are *protected* data members. This means that they can be accessed freely by descendants of the class.

```
int num_inputs;
int num_outputs;
float *outputs;  // pointer to array of outputs
float *inputs;   // pointer to array of inputs, which
                 // are outputs of some other layer
friend network;
```

There are also two pointers to arrays of floats in this class. They are the pointers to the outputs in a given layer and the inputs to a given layer. To get a better idea of what a layer encompasses, Figure 7.3 shows you a small backpropagation network, with a dotted line that shows you the three layers for that network. A layer contains neurons and weights. The layer is responsible for calculating its output (**calc_out()**), stored in the **float * outputs array**, and errors (**calc_error()**) for each of its respective neurons. The errors are stored in another array called **float * output_errors** defined in the **output** class. Note that the **input** class does not have any weights associated with it and therefore is a special case. It does not need to provide any data members or function members related to errors or backpropagation. The only purpose of the input layer is to store data to be forward propagated to the next layer.

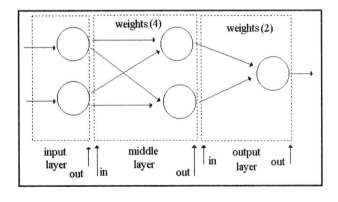

FIGURE 7.3 *Organization of layers for backpropagation program.*

With the output layer, there are a few more arrays present. First, for storing backpropagated errors, there is an array called **float * back_errors**. There is a weights array called **float * weights**, and finally, for storing the expected values that initiate the error calculation process, there is an array called **float * expected_values**. Note that the middle layer needs almost all of these arrays and inherits them by being a derived class of the **output_layer** class.

There is one other class besides the **layer** class and its descendants defined in this header file, and that is the **network** class, which is used to set up communication channels between layers and to feed and remove data from the network. The **network** class performs the interconnection of layers by setting the pointer of an input array of a given layer to the output array of a previous layer.

> This is a fairly extensible scheme that can be used to create variations on the backpropagation network with feedback connections, for instance.
>
> N O T E

Another connection that the network class is responsible for is setting the pointer of an **output_error** array to the **back_error** array of the next layer (remember errors flow in reverse, and the **back_error** array is the output error of the layer reflected at its inputs).

The **network** class stores an array of pointers to layers and an array of layer sizes for all the layers defined. These layer objects and arrays are dynamically allocated on the heap with the **New** and **Delete** functions in C++. There is some minimal error checking for file I/O and memory allocation, which can be enhanced, if desired.

As you can see, the backpropagation network can quickly become a memory and CPU hog, with large networks and large training sets. The size and topology of the network, or *architecture*, will largely dictate both these characteristics.

Details of the Backpropagation Implementation File

The implementation of the classes and *methods* is the next topic. Let's look at the layer.cpp file in Listing 7.2. Following the listing will be a summary of the important functions and their purpose.

Listing 7.2 layer.cpp implementation file for the backpropagation simulator

```cpp
// layer.cpp       V.Rao, H.Rao
// compile for floating point hardware if available
#include <stdio.h>
#include <iostream.h>
#include <stdlib.h>
#include <math.h>
#include <time.h>
#include "layer.h"


inline float squash(float input)
// squashing function
// use sigmoid -- can customize to something
// else if desired; can add a bias term too
//
{
if (input < -50)
    return 0.0;
else   if (input > 50)
        return 1.0;
    else return (float)(1/(1+exp(-(double)input)));

}

inline float randomweight(unsigned init)
{
int num;
// random number generator
// will return a floating point
// value between -1 and 1

if (init==1) // seed the generator
    srand ((unsigned)time(NULL));

num=rand() % 100;

return 2*(float(num/100.00))-1;
}

// the next function is needed for Turbo C++
// and Borland C++ to link in the appropriate
// functions for fscanf floating point formats:
static void force_fpf()
{
    float x, *y;
    y=&x;
    x=*y;
}
```

```
// -------------------------------------------
//                 input layer
//-------------------------------------------
input_layer::input_layer(int i, int o)
{

num_inputs=i;
num_outputs=o;

outputs = new float[num_outputs];
if (outputs==0)
    {
    cout << "not enough memory\n";
    cout << "choose a smaller architecture\n";
    exit(1);
    }
}

input_layer::~input_layer()
{
delete [num_outputs] outputs;
}


void input_layer::calc_out()
{
//nothing to do, yet
}

// -------------------------------------------
//                 output layer
//-------------------------------------------

output_layer::output_layer(int i, int o)
{

num_inputs=i;
num_outputs=o;
weights = new float[num_inputs*num_outputs];
output_errors = new float[num_outputs];
back_errors = new float[num_inputs];
outputs = new float[num_outputs];
expected_values = new float[num_outputs];
if ((weights==0)||(output_errors==0)||(back_errors==0)
    ||(outputs==0)||(expected_values==0))
    {
    cout << "not enough memory\n";
    cout << "choose a smaller architecture\n";
    exit(1);
    }
```

```
}

output_layer::~output_layer()
{
// some compilers may require the array
// size in the delete statement; those
// conforming to Ansi C++ will not
delete [num_outputs*num_inputs] weights;
delete [num_outputs] output_errors;
delete [num_inputs] back_errors;
delete [num_outputs] outputs;

}


void output_layer::calc_out()
{

int i,j,k;
float accumulator=0.0;


for (j=0; j<num_outputs; j++)
    {

    for (i=0; i<num_inputs; i++)

        {
        k=i*num_outputs;
        if (weights[k+j]*weights[k+j] > 1000000.0)
            {
            cout << "weights are blowing up\n";
            cout << "try a smaller learning constant\n";
            cout << "e.g. beta=0.02    aborting...\n";
            exit(1);
            }
        outputs[j]=weights[k+j]*(*(inputs+i));
        accumulator+=outputs[j];
        }
    // use the sigmoid squash function
    outputs[j]=squash(accumulator);
    accumulator=0;
    }

}



void output_layer::calc_error(float & error)
{
int i, j, k;
```

```
float accumulator=0;
float total_error=0;

for (j=0; j<num_outputs; j++)
    {
    output_errors[j] = expected_values[j]-outputs[j];
    total_error+=output_errors[j];
    }

error=total_error;

for (i=0; i<num_inputs; i++)
    {
    k=i*num_outputs;
    for (j=0; j<num_outputs; j++)
        {
        back_errors[i]=
            weights[k+j]*output_errors[j];
        accumulator+=back_errors[i];
        }
    back_errors[i]=accumulator;
    accumulator=0;
    // now multiply by derivative of
    // sigmoid squashing function, which is
    // just the input*(1-input)
    back_errors[i]*=(*(inputs+i))*(1-(*(inputs+i)));
    }

}

void output_layer::randomize_weights()
{
int i, j, k;
const unsigned first_time=1;

const unsigned not_first_time=0;
float discard;

discard=randomweight(first_time);

for (i=0; i< num_inputs; i++)
    {
    k=i*num_outputs;
    for (j=0; j< num_outputs; j++)
        weights[k+j]=randomweight(not_first_time);
    }
}

void output_layer::update_weights(const float beta)
{
int i, j, k;
```

```
// learning law: weight_change =
//     beta*output_error*input

for (i=0; i< num_inputs; i++)
    {
    k=i*num_outputs;
    for (j=0; j< num_outputs; j++)
        weights[k+j] +=
            beta*output_errors[i]*(*(inputs+i));
    }

}

void output_layer::list_weights()
{
int i, j, k;

for (i=0; i< num_inputs; i++)
    {
    k=i*num_outputs;
    for (j=0; j< num_outputs; j++)
        cout << "weight["<<i<<","<<
            j<<"] is: "<<weights[k+j];
    }

}

void output_layer::list_errors()
{
int i, j;

for (i=0; i< num_inputs; i++)
    cout << "backerror["<<i<<
        "] is : "<<back_errors[i]<<"\n";

for (j=0; j< num_outputs; j++)
    cout << "outputerrors["<<j<<
        "] is: "<<output_errors[j]<<"\n";

}


void output_layer::write_weights(int layer_no,
        FILE * weights_file_ptr)
{
int i, j, k;

// assume file is already open and ready for
// writing
```

```
// prepend the layer_no to all lines of data
// format:
//     layer_no  weight[0,0] weight[0,1] ...
//     layer_no  weight[1,0] weight[1,1] ...
//     ...

for (i=0; i< num_inputs; i++)
    {
    fprintf(weights_file_ptr,"%i ",layer_no);
    k=i*num_outputs;
     for (j=0; j< num_outputs; j++)
        {
        fprintf(weights_file_ptr,"%f",
            weights[k+j]);
        }
    fprintf(weights_file_ptr,"\n");
    }

}

void output_layer::read_weights(int layer_no,
        FILE * weights_file_ptr)
{
int i, j, k;

// assume file is already open and ready for
// reading

// look for the prepended layer_no
// format:
//     layer_no  weight[0,0] weight[0,1] ...
//     layer_no  weight[1,0] weight[1,1] ...
//     ...
while (1)

    {

    fscanf(weights_file_ptr,"%i",&j);
    if ((j==layer_no)|| (feof(weights_file_ptr)))
        break;
    else
        {
        while (fgetc(weights_file_ptr) != '\n')
            {;}// get rest of line
        }
    }

if (!(feof(weights_file_ptr)))
    {
    // continue getting first line
    i=0;
```

```
        for (j=0; j< num_outputs; j++)
            {

            fscanf(weights_file_ptr,"%f",
                    &weights[j]); // i*num_outputs = 0
            }
        fscanf(weights_file_ptr,"\n");

        // now get the other lines
        for (i=1; i< num_inputs; i++)
            {
            fscanf(weights_file_ptr,"%i",&layer_no);
            k=i*num_outputs;
            for (j=0; j< num_outputs; j++)
                {
                fscanf(weights_file_ptr,"%f",
                    &weights[k+j]);
                }

            }
        fscanf(weights_file_ptr,"\n");
        }

else cout << "end of file reached\n";

}
void output_layer::list_outputs()
{
int j;

for (j=0; j< num_outputs; j++)
    {
    cout << "outputs["<<j
        <<"] is: "<<outputs[j]<<"\n";
    }

}

// -----------------------------------------
//              middle layer
//------------------------------------------

middle_layer::middle_layer(int i, int o):
    output_layer(i,o)
{

}

middle_layer::~middle_layer()
{
```

```
delete [num_outputs*num_inputs] weights;
delete [num_outputs] output_errors;
delete [num_inputs] back_errors;
delete [num_outputs] outputs;
}

void middle_layer::calc_error()
{
int i, j, k;
float accumulator=0;

for (i=0; i<num_inputs; i++)
    {
    k=i*num_outputs;
    for (j=0; j<num_outputs; j++)
        {
        back_errors[i]=
            weights[k+j]*(*(output_errors+j));
        accumulator+=back_errors[i];
        }
    back_errors[i]=accumulator;
    accumulator=0;
    // now multiply by derivative of
    // sigmoid squashing function, which is
    // just the input*(1-input)
    back_errors[i]*=(*(inputs+i))*(1-(*(inputs+i)));
    }

}

network::network()
{
position=0L;
}

network::~network()
{
int i,j,k;
i=layer_ptr[0]->num_outputs;// inputs
j=layer_ptr[number_of_layers-1]->num_outputs; //outputs
k=MAX_VECTORS;


delete [(i+j)*k]buffer;
}

void network::set_training(const unsigned & value)
{
training=value;
}
```

```
unsigned network::get_training_value()
{
return training;
}

void network::get_layer_info()
{
int i;

//--------------------------------------------
//
// Get layer sizes for the network
//
// ----------------------------------------

cout << " Please enter in the number of layers for your net-
work.\n";
cout << " You can have a minimum of 3 to a maximum of 5. \n";
cout << " 3 implies 1 hidden layer; 5 implies 3 hidden layers
: \n\n";

cin >> number_of_layers;

cout << " Enter in the layer sizes separated by spaces.\n";
cout << " For a network with 3 neurons in the input \
layer,\n";
cout << " 2 neurons in a hidden layer, and 4 neurons in \
the\n";
cout << " output layer, you would enter: 3 2 4 .\n";
cout << " You can have up to 3 hidden layers,for five maxi-\
mum entries :\n\n";

for (i=0; i<number_of_layers; i++)
    {
    cin >> layer_size[i];
    }

// -----------------------------------------------------------
// size of layers:
//     input_layer        layer_size[0]
//     output_layer       layer_size[number_of_layers-1]
//     middle_layers      layer_size[1]
//               optional: layer_size[number_of_layers-3]
//               optional: layer_size[number_of_layers-2]
//------------------------------------------------------------

}

void network::set_up_network()
{
int i,j,k;
```

```
//-----------------------------------------------------------
// Construct the layers
//
//-----------------------------------------------------------

layer_ptr[0] = new input_layer(0,layer_size[0]);

for (i=0;i<(number_of_layers-1);i++)
    {
    layer_ptr[i+1] =
    new middle_layer(layer_size[i],layer_size[i+1]);
    }

layer_ptr[number_of_layers-1] = new
output_layer(layer_size[number_of_layers-2],layer_size[num-
ber_of_layers-1]);

for (i=0;i<(number_of_layers-1);i++)
    {
    if (layer_ptr[i] == 0)
        {
        cout << "insufficient memory\n";
        cout << "use a smaller architecture\n";
        exit(1);
        }
    }

//-----------------------------------------------------------
// Connect the layers
//
//-----------------------------------------------------------
// set inputs to previous layer outputs for all layers,
// except the input layer

for (i=1; i< number_of_layers; i++)
    layer_ptr[i]->inputs = layer_ptr[i-1]->outputs;

// for back_propagation, set output_errors to next layer
//      back_errors for all layers except the output
//      layer and input layer


for (i=1; i< number_of_layers -1; i++)
    ((output_layer *)layer_ptr[i])->output_errors =
        ((output_layer *)layer_ptr[i+1])->back_errors;

// define the IObuffer that caches data from
// the datafile
i=layer_ptr[0]->num_outputs;// inputs
j=layer_ptr[number_of_layers-1]->num_outputs; //outputs
k=MAX_VECTORS;
```

```
buffer=new
    float[(i+j)*k];
if (buffer==0)
    cout << "insufficient memory for buffer\n";
}

void network::randomize_weights()
{
int i;

for (i=1; i<number_of_layers; i++)
    ((output_layer *)layer_ptr[i])
        ->randomize_weights();
}


void network::update_weights(const float beta)
{
int i;

for (i=1; i<number_of_layers; i++)
    ((output_layer *)layer_ptr[i])
        ->update_weights(beta);
}

void network::write_weights(FILE * weights_file_ptr)
{
int i;

for (i=1; i<number_of_layers; i++)
    ((output_layer *)layer_ptr[i])
        ->write_weights(i,weights_file_ptr);
}

void network::read_weights(FILE * weights_file_ptr)
{
int i;

for (i=1; i<number_of_layers; i++)
    ((output_layer *)layer_ptr[i])
        ->read_weights(i,weights_file_ptr);
}

void network::list_weights()
{
int i;

for (i=1; i<number_of_layers; i++)
    {
    cout << "layer number : " <<i<< "\n";
```

```
    ((output_layer *)layer_ptr[i])
        ->list_weights();
    }
}

void network::list_outputs()
{
int i;

for (i=1; i<number_of_layers; i++)
    {
    cout << "layer number : " <<i<< "\n";
    ((output_layer *)layer_ptr[i])
        ->list_outputs();
    }
}

void network::write_outputs(FILE *outfile)
{
int i, ins, outs;
ins=layer_ptr[0]->num_outputs;
outs=layer_ptr[number_of_layers-1]->num_outputs;
float temp;

fprintf(outfile,"for input vector:\n");

for (i=0; i<ins; i++)
    {
    temp=layer_ptr[0]->outputs[i];
    fprintf(outfile,"%f  ",temp);
    }

fprintf(outfile,"\noutput vector is:\n");

for (i=0; i<outs; i++)
    {
    temp=layer_ptr[number_of_layers-1]->
    outputs[i];
    fprintf(outfile,"%f  ",temp);

    }

if (training==1)
{
fprintf(outfile,"\nexpected output vector is:\n");

for (i=0; i<outs; i++)
    {
    temp=((output_layer *)(layer_ptr[number_of_layers-1]))->
    expected_values[i];
    fprintf(outfile,"%f  ",temp);
```

```
        }
    }

    fprintf(outfile,"\n--------------------\n");

    }

    void network::list_errors()
    {
    int i;

    for (i=1; i<number_of_layers; i++)
        {
        cout << "layer number : " <<i<< "\n";
        ((output_layer *)layer_ptr[i])
            ->list_errors();
        }
    }

    int network::fill_IObuffer(FILE * inputfile)
    {
    // this routine fills memory with
    // an array of input, output vectors
    // up to a maximum capacity of
    // MAX_INPUT_VECTORS_IN_ARRAY
    // the return value is the number of read
    // vectors


    int i, k, count, veclength;

    int ins, outs;

    ins=layer_ptr[0]->num_outputs;

    outs=layer_ptr[number_of_layers-1]->num_outputs;

    if (training==1)
        veclength=ins+outs;
    else
        veclength=ins;

    count=0;
    while  ((count<MAX_VECTORS)&&
            (!feof(inputfile)))
        {
        k=count*(veclength);
        for (i=0; i<veclength; i++)
            {
            fscanf(inputfile,"%f",&buffer[k+i]);
```

```
            }
        fscanf(inputfile,"\n");
        count++;
        }

    if (!(ferror(inputfile)))
        return count;
    else return -1; // error condition

    }

    void network::set_up_pattern(int buffer_index)
    {
    // read one vector into the network
    int i, k;
    int ins, outs;

    ins=layer_ptr[0]->num_outputs;
    outs=layer_ptr[number_of_layers-1]->num_outputs;
    if (training==1)
        k=buffer_index*(ins+outs);
    else
        k=buffer_index*ins;

    for (i=0; i<ins; i++)
        layer_ptr[0]->outputs[i]=buffer[k+i];

    if (training==1)
    {
        for (i=0; i<outs; i++)

            ((output_layer *)layer_ptr[number_of_layers-1])->
                expected_values[i]=buffer[k+i+ins];
    }

    }


    void network::forward_prop()
    {
    int i;
    for (i=0; i<number_of_layers; i++)
        {
        layer_ptr[i]->calc_out(); //polymorphic
                    // function
        }
    }

    void network::backward_prop(float & toterror)
    {
```

```
int i;

// error for the output layer
((output_layer*)layer_ptr[number_of_layers-1])->
        calc_error(toterror);

// error for the middle layer(s)
for (i=number_of_layers-2; i>0; i--)
    {
    ((middle_layer*)layer_ptr[i])->
        calc_error();

    }

}
```

A Look at the Functions in the layer.cpp File

◆ **void set_training(const unsigned &)**   Sets the value of the private data member, training; use 1 for training mode, and 0 for test mode

◆ **unsigned get_training_value()**   Gets the value of the training constant that gives the mode in use

◆ **void get_layer_info()**   Gets information about the number of layers and layer sizes from the user

◆ **void set_up_network()**   This routine sets up the connections between layers by assigning pointers appropriately

◆ **void randomize_weights()**   At the beginning of the training process, this routine is used to randomize all of the weights in the network

◆ **void update_weights(const float)**   As part of training, weights are updated according to the learning law used in backpropagation

◆ **void write_weights(FILE *)**   This routine is used to write weights to a file

◆ **void read_weights(FILE *)**   This routine is used to read weights into the network from a file

◆ **void list_weights()**   This routine can be used to list weights while a simulation is in progress

◆ **void write_outputs(FILE *)**   This routine writes the outputs of the network to a file

◆ **void list_outputs()**   This routine can be used to list the outputs of the network while a simulation is in progress

◆ **void list_errors()**   Lists errors for all layers while a simulation is in progress

◆ **void forward_prop()**   Performs the forward propagation

◆ **void backward_prop(float &)**   Pperforms the backward error propagation

◆ **int fill_IObuffer(FILE *)**   This routine fills the internal IO buffer with data from the training or test data sets

◆ **void set_up_pattern(int)**   This routine is used to set up one pattern from the IO buffer for training

◆ **inline float squash(float input)**   This function performs the sigmoid function

◆ **inline float randomweight (unsigned unit)**   This routine returns a random weight between –1 and 1; use 1 to initialize the generator, and 0 for all subsequent calls

Note that the functions **squash(float)** and **randomweight(unsigned)** are declared **inline**. This means that the function's source code is inserted wherever it appears. This increases code size, but increases speed because a function call, which is expensive, is avoided.

N O T E

The final file to look at is the backprop.cpp file presented in Listing 7.3 .

Listing 7.3   The backprop.cpp file for the backpropagation simulator

```
// backprop.cpp      V. Rao, H. Rao
#include "layer.cpp"

#define TRAINING_FILE   "training.dat"
#define WEIGHTS_FILE "weights.dat"
#define OUTPUT_FILE "output.dat"
#define TEST_FILE    "test.dat"

void main()
{

float error_tolerance=0.1;
float total_error=0.0;
float avg_error_per_cycle=0.0;
```

```
float error_last_cycle=0.0;
float avgerr_per_pattern=0.0; // for the latest cycle
float error_last_pattern=0.0;
float learning_parameter=0.02;
unsigned temp, startup;
long int vectors_in_buffer;
long int max_cycles;
long int patterns_per_cycle=0;

long int total_cycles, total_patterns;
int i;

// create a network object
network backp;

FILE * training_file_ptr, * weights_file_ptr, * output_file_ptr;
FILE * test_file_ptr, * data_file_ptr;

// open output file for writing
if ((output_file_ptr=fopen(OUTPUT_FILE,"w"))==NULL)
    {
    cout << "problem opening output file\n";
    exit(1);
    }

// enter the training mode : 1=training on      0=training off
cout << "-------------------------------------------------\n";
cout << " C++ Neural Networks and Fuzzy Logic \n";
cout << " Backpropagation simulator \n";
cout << "     version 1 \n";
cout << "-------------------------------------------------\n";
cout << "Please enter 1 for TRAINING on, or 0 for off: \n\n";
cout << "Use training to change weights according to your\n";
cout << "expected outputs. Your training.dat file should \
    contain\n";
cout << "a set of inputs and expected outputs. The number \
    of\n";
cout << "inputs determines the size of the first (input) \
    layer\n";
cout << "while the number of outputs determines the size of \
    the\n";
cout << "last (output) layer :\n\n";

cin >> temp;
backp.set_training(temp);

if (backp.get_training_value() == 1)
    {
```

```
        cout << "--> Training mode is *ON*. weights will be \
        saved\n";
        cout << "in the file weights.dat at the end of the\n";
        cout << "current set of input (training) data\n";
        }
    else
        {
        cout << "--> Training mode is *OFF*. weights will be \
            loaded\n";
        cout << "from the file weights.dat and the current\n";
        cout << "(test) data set will be used. For the test\n";
        cout << "data set, the test.dat file should contain\n";
        cout << "only inputs, and no expected outputs.\n";
        }

    if (backp.get_training_value()==1)
        {
        // -----------------------------------------
        // Read in values for the error_tolerance,
        // and the learning_parameter
        // -----------------------------------------
        cout << " Please enter in the error_tolerance\n";
        cout << " --- between 0.001 to 100.0, try 0.1 to start \
        \n";
        cout << "\n";
        cout << "and the learning_parameter, beta\n";
        cout << " --- between 0.01 to 1.0, try 0.5 to start -- \
        \n\n";
        cout << " separate entries by a space\n";
        cout << " example: 0.1 0.5 sets defaults mentioned :\n\n";

        cin >> error_tolerance >> learning_parameter;
        //------------------------------------------------
        // open training file for reading
        //------------------------------------------------
        if ((training_file_ptr=fopen(TRAINING_FILE,"r"))==NULL)
            {
            cout << "problem opening training file\n";
            exit(1);
            }
        data_file_ptr=training_file_ptr; // training on

        // Read in the maximum number of cycles
        // each pass through the input data file is a cycle
        cout << "Please enter the maximum cycles for the simula-\
        tion\n";
        cout << "A cycle is one pass through the data set.\n";
        cout << "Try a value of 10 to start with\n";

        cin >> max_cycles;
```

```
        }
    else
        {
        if ((test_file_ptr=fopen(TEST_FILE,"r"))==NULL)
            {
            cout << "problem opening test file\n";
            exit(1);
            }

        data_file_ptr=test_file_ptr; // training off
        }



//
// training: continue looping until the total error is less
// than
//      the tolerance specified, or the maximum number of
//      cycles is exceeded; use both the forward signal
//      propagation
//      and the backward error propagation phases. If the error
//      tolerance criteria is satisfied, save the weights in a
//      file.
// no training: just proceed through the input data set once
//      in the
//      forward signal propagation phase only. Read the
//      starting
//      weights from a file.
// in both cases report the outputs on the screen


// initialize counters
total_cycles=0; // a cycle is once through all the input data
total_patterns=0; // a pattern is one entry in the input data



// get layer information
backp.get_layer_info();

// set up the network connections
backp.set_up_network();

// initialize the weights
if (backp.get_training_value()==1)
    {
    // randomize weights for all layers; there is no
    // weight matrix associated with the input layer
    // weight file will be written after processing
```

```
    // so open for writing
    if ((weights_file_ptr=fopen(WEIGHTS_FILE,"w"))
        ==NULL)
        {
        cout << "problem opening weights file\n";
        exit(1);
        }
    backp.randomize_weights();
    }
else
    {
    // read in the weight matrix defined by a
    // prior run of the backpropagation simulator
    // with training on
    if ((weights_file_ptr=fopen(WEIGHTS_FILE,"r"))
        ==NULL)
        {
        cout << "problem opening weights file\n";
        exit(1);
        }
    backp.read_weights(weights_file_ptr);
    }



// main loop
// if training is on, keep going through the input data
//     until the error is acceptable or the maximum number of
//     cycles
//     is exceeded.
// if training is off, go through the input data once. report
// outputs
// with inputs to file output.dat

startup=1;
vectors_in_buffer = MAX_VECTORS; // startup condition
total_error = 0;

while (            ((backp.get_training_value()==1)
        && (avgerr_per_pattern
            > error_tolerance)
        && (total_cycles < max_cycles)
        && (vectors_in_buffer !=0))
        || ((backp.get_training_value()==0)
        && (total_cycles < 1))
        || ((backp.get_training_value()==1)
        && (startup==1))
        )
{
startup=0;
error_last_cycle=0; // reset for each cycle
```

```
    patterns_per_cycle=0;
    // process all the vectors in the datafile
    // going through one buffer at a time
    // pattern by pattern


    while ((vectors_in_buffer==MAX_VECTORS))
        {

        vectors_in_buffer=
            backp.fill_IObuffer(data_file_ptr); // fill buffer
            if (vectors_in_buffer < 0)
                {
                cout << "error in reading in vectors, aborting\n";
                cout << "check that there are on extra \
                linefeeds\n";
                cout << "in your data file, and that the number\n";
                cout << "of layers and size of layers match the\n";
                cout << "the parameters provided.\n";
                exit(1);
                }

            // process vectors
            for (i=0; i<vectors_in_buffer; i++)
                {
                // get next pattern
                backp.set_up_pattern(i);

                total_patterns++;
                patterns_per_cycle++;
                // forward propagate

                backp.forward_prop();

                if (backp.get_training_value()==0)
                    backp.write_outputs(output_file_ptr);

                // back_propagate, if appropriate
                if (backp.get_training_value()==1)
                    {

                    backp.backward_prop(error_last_pattern);
                    error_last_cycle += error_last_pattern
                    *error_last_pattern;
                    backp.update_weights(learning_parameter);
                    // backp.list_weights(); // can
                    // see change in weights by
                    // using list_weights before and
                    // after back_propagation
                    }
```

```
        }

    error_last_pattern = 0;
        }

avgerr_per_pattern=((float)sqrt((double)error_last_cycle))
/patterns_per_cycle;
total_error += error_last_cycle;
total_cycles++;

// most character displays are 26 lines
// user will see a corner display of the cycle count
// as it changes

cout << "\n\n\n\n\n\n\n\n\n\n\n\n\n\n\n\n\n\n\n\n\n\n\n\n\n";
cout << total_cycles << "\t" << avgerr_per_pattern << "\n";

fseek(data_file_ptr, 0L, SEEK_SET); // reset the file pointer
            // to the beginning of
            // the file
vectors_in_buffer = MAX_VECTORS; // reset

} // end main loop

cout << "\n\n\n\n\n\n\n\n\n\n\n";
cout << "-------------------------------------------------\n";
cout << " done:  results in file output.dat\n";
cout << "     training: last vector only\n";
cout << "     not training: full cycle\n\n";
if (backp.get_training_value()==1)
    {
    backp.write_weights(weights_file_ptr);
    backp.write_outputs(output_file_ptr);
    avg_error_per_cycle = (float)sqrt((double)total_error)/
    total_cycles;
    error_last_cycle = (float)sqrt((double)error_last_cycle);


cout << "     weights saved in file weights.dat\n";
cout << "\n";
cout << "---->average error per cycle = " <<
    avg_error_per_cycle << " <---\n";
cout << "-->error last cycle= " << error_last_cycle << " <--\
    \n";
cout << "->error last cycle per pattern= " << avgerr_per_pat\
    tern << " <---\n";

    }
```

```
cout << "----------->total cycles = " << total_cycles <<
" <---\n";
cout << "----------->total patterns = " << total_patterns <<
" <---\n";
cout << "---------------------------------------------\n";
// close all files
fclose(data_file_ptr);
fclose(weights_file_ptr);
fclose(output_file_ptr);

}
```

The backprop.cpp file implements the simulator controls. First, data is accepted from the user for network parameters. Assuming **Training** mode is used, the training file is opened and data is read from the file to fill the IO buffer. Then the main loop is executed where the network processes pattern by pattern to complete a cycle, which is one pass through the entire training data set. (The IO buffer is refilled as required during this process.) After executing one cycle, the file pointer is reset to the beginning of the file and another cycle begins. The simulator continues with cycles until one of the two fundamental criteria is met:

1.   The maximum cycle count specified by the user is exceeded.
2.   The average error per pattern for the latest cycle is less than the error tolerance specified by the user.

When either of these occurs, the simulator stops and reports out the error achieved, and saves weights in the weights.dat file and one output vector in the output.dat file.

In the case of **Test** mode, exactly one cycle is processed by the network, and outputs are written to the output.dat file. At the beginning of the simulation in **Test** mode, the network is set up with weights from the weights.dat file. To simplify the program, the user is requested to enter the number of layers and size of layers, although you could have the program figure this out from the weights file.

Compiling and Running the Backpropagation Simulator

Compiling the backprop.cpp file will compile the simulator since layer.cpp is included in backprop.cpp. To run the simulator, once you have created

an executable (using 80 x 87 floating point hardware if available), you type
in backprop and see the following screen (computer output in boldface):

C++ Neural Networks and Fuzzy Logic
Backpropagation simulator
version 1
Please enter 1 for TRAINING on, or 0 for off:

Use training to change weights according to your
expected outputs. Your training.dat file should contain
a set of inputs and expected outputs. The number of
inputs determines the size of the first (input) layer
while the number of outputs determines the size of the
last (output) layer :

1

--> Training mode is *ON*. weights will be saved
in the file weights.dat at the end of the
current set of input (training) data
 Please enter in the error_tolerance
--- between 0.001 to 100.0, try 0.1 to start --

and the learning_parameter, beta
--- between 0.01 to 1.0, try 0.5 to start --

 separate entries by a space
 example: 0.1 0.5 sets defaults mentioned :

0.2 0.25

Please enter the maximum cycles for the simulation
A cycle is one pass through the data set.
Try a value of 10 to start with
Please enter in the number of layers for your network.
You can have a minimum of three to a maximum of five.
three implies one hidden layer; five implies three hidden layers:

3

Enter in the layer sizes separated by spaces.
For a network with three neurons in the input layer,
two neurons in a hidden layer, and four neurons in the
output layer, you would enter: 3 2 4.
You can have up to three hidden layers,for five maximum entries :

2 2 1

1   0.353248
2   0.352684

```
3    0.352113
4    0.351536
5    0.350954
...
299                0.0582381
300                0.0577085
```
- -

done: results in file output.dat
training: last vector only
not training: full cycle

weights saved in file weights.dat

---->average error per cycle = 0.20268 <---
---->error last cycle = 0.0577085 <---
->error last cycle per pattern= 0.0577085 <---
------------>total cycles = 300 <---
------------>total patterns = 300 <---

N O T E

The cycle number and the average error per pattern is displayed as the simulation progresses (not all values shown). You can monitor this to make sure the simulator is converging on a solution. If the error does not seem to decrease beyond a certain point, but instead drifts or blows up, then you should start the simulator again with a new starting point defined by the random weights initializer. Also, you could try decreasing the size of the learning rate parameter. Learning may be slower, but this may allow a better minimum to be found.

This example shows just one pattern in the training set with two inputs and one output. The results along with the (one) last pattern are shown as follows from the file output.dat:

```
for input vector:
0.400000  -0.400000
output vector is:
0.842291
expected output vector is:
0.900000
```

The match is pretty good as can be expected, since the optimization is easy for the network; there is only one pattern to worry about. Let's look at the final set of weights for this simulation in weights.dat:

```
1 0.1750390.435039
```

```
1 -1.319244-0.559244
2 0.358281
2 2.421172
```

These are the weights that were obtained by the updating the weights for 300 cycles with the learning law.

N O T E

We'll leave the backpropagation simulator for now and return to it in a later chapter for further exploration. You can experiment a number of different ways with the simulator:

♦ Try a different number of layers and layer sizes for a given problem.

♦ Try different learning rate parameters and see its effect on convergence and training time.

♦ Try a very large learning rate parameter (should be between 0 and 1); try a number over 1 and note the result.

Summary

In this chapter you are acquainted with one of the most powerful neural network algorithms called backpropagation. Without having feedback connections, propagating only errors appropriately to the hidden layer and input layer connections, the algorithm uses the so-called generalized delta rule and trains the network with exemplar pairs of patterns. It is a difficult task to determine how many hidden-layer neurons are to be provided for. The number of hidden layers could be more than one. In general, the size of the hidden layer(s) is related to the features or distinguishing characteristics that should be discerned from the data. Our example in this chapter relates to a simple case where there is a single hidden layer. The outputs of the output neurons, and therefore of the network, are vectors with components between 0 and 1, since the **thresholding** function is the **sigmoid** function. These values can be scaled, if necessary, to get values in another interval.

Our example does not relate to any particular function to be computed by the network, but inputs and outputs were randomly chosen. What this can tell you is that, if you do not know the functional equation between two sets of vectors, the backpropagation network can find the mapping for any vector in the domain, even if the functional equation is not found. For all we know, that function could be nonlinear as well.

NOTE

There is one important fact you need to remember about the backpropagation algorithm. Its steepest descent procedure in training does not guarantee finding a global or overall minimum, but it can find only a local minimum of the energy surface.

Chapter 8

BAM: Bidirectional Associative Memory

BAM (Bidirectional Associative Memory) Model

The Bidirectional Associative Memory model has a neural network of two layers and is fully connected from each layer to the other. That is, there are feedback connections from the output layer to the input layer. However, the weights on the connections between any two given neurons from different layers are the same. You may even consider it to be a single bidirectional connection with a single weight. The matrix of weights for the connections from the output layer to the input layer is simply the transpose of the matrix of weights for the connections between the input and output layer. If we denote the matrix for forward connection weights by $\mathbf{W}$, then $\mathbf{W^T}$ gives the matrix of weight for the output layer to input layer connections. As you recall, the transpose of a matrix is obtained simply by interchanging the rows and the columns of the matrix.

There are two layers of neurons, an input layer and an output layer. There are no lateral connections, that is, no two neurons within the same layer are connected. *Recurrent* connections, which are feedback connections to a neuron from itself, may or may not be present. The architecture is quite simple. Figure 8.1 shows the layout for this neural network model, using only three input neurons and two output neurons. There are feedback connections. This figure also indicates the presence of inputs and outputs at each of the two fields for the bidirectional associative memory network. Connection weights are also shown as labels on only a few connections in this figure, to avoid cluttering. The general case is analogous.

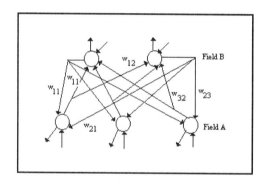

FIGURE 8.1   *Layout of a BAM network*

Inputs and Outputs

The input to a BAM network is a vector of real numbers, usually in the set { –1, +1 }. The output is also a vector of real numbers, usually in the set { –1, +1 }, with the same or different dimension from that of the input. These vectors can be considered as patterns, and the network makes heteroassociation of patterns. If the output is required to be the same as input, then you are asking the network to make autoassociation, which it does, and it becomes a special case of the general activity of this type of neural network.

For inputs and outputs that are not over the set containing just –1 and +1, try following this next procedure. You can first make a mapping into binary numbers, and then a mapping of each binary digit into a bipolar digit. For example, if your inputs are first names of people, each character in the name can be replaced by its ASCII code, which in turn can be changed to a binary number, and then each binary digit 0 can be replaced by –1. For example, the ASCII code for the letter R is 82, which is 1010010, as a binary number. This is mapped on to the bipolar string 1 –1 1 –1 –1 1 –1. If a name consists of three characters, their ASCII codes in binary can be concatenated or juxtaposed and the corresponding bipolar string obtained. This bipolar string can also be looked upon as a vector of bipolar characters.

Weights and Training

BAM does not modify weights during its operation, which indicates a lack of training. The adaptive variety of BAM, called the *Adaptive Bidirectional Associative Memory*, (ABAM), does undergo training. Yet BAM needs some exemplar pairs of vectors. The pairs used as exemplars are those that possess heteroassociation. The weight matrix, there are two, but one is just the transpose of the other as already mentioned, is constructed in terms of the exemplar vector pairs.

The use of exemplar vectors, you may say, is an off-line learning—to determine what the weights should be. Once weights are so determined, and an input vector is presented, a potentially associated vector is outputted. It is taken as input in the opposite direction, and its potentially associated vector is obtained back at the input layer. If the last vector found is the same as what is originally input, then there is *resonance*. Suppose the vector *B* is obtained at one end, as a result of *C* being input at the other end. Then if *B* in turn is input during the next cycle of operation at the end where it was obtained, and produces *C* at the opposite end, then you have a pair of heteroassociated vectors. This is what is basically happening in a BAM neural network.

What follow are the equations for the determination of the weight matrix, when the k pairs of exemplar vectors are denoted by (X_i, Y_i), i ranging from 1 to k. Note that T in the superscript of a matrix stands for the transpose of the matrix. Whereas you interchange the rows and columns to get the transpose of a matrix, you write a column vector as a row vector, and vice versa to get the transpose of a vector. The following equations refer to the vector pairs after their components are changed to bipolar values, only for obtaining the weight matrix **W**. Once **W** is obtained, further use of these exemplar vectors is made in their original form.

$$W = X_1{}^T Y_1 + \ldots + X_k{}^T Y_k$$

and

$$W^T = Y_1{}^T X_1 + \ldots + Y_k{}^T X_k$$

Example

Suppose you choose two pairs of vectors as possible exemplars. Let them be,

$$X_1 = (1, 0, 0, 1), Y_1 = (0, 1, 1)$$
and
$$X_2 = (0, 1, 1, 0), Y_2 = (1, 0, 1)$$

These you change into bipolar components and get, respectively, $(1, -1, -1, 1)$, $(-1, 1, 1)$, $(-1, 1, 1, -1)$, and $(1, -1, 1)$.

$$
W = \begin{matrix} 1 \\ -1 \\ -1 \\ 1 \end{matrix} \begin{bmatrix} -1 & 1 & 1 \end{bmatrix} + \begin{matrix} -1 \\ 1 \\ 1 \\ -1 \end{matrix} \begin{bmatrix} 1 & -1 & 1 \end{bmatrix} = \begin{matrix} -1 & 1 & 1 \\ 1 & -1 & -1 \\ 1 & -1 & -1 \\ -1 & 1 & 1 \end{matrix} + \begin{matrix} -1 & 1 & -1 \\ 1 & -1 & 1 \\ 1 & -1 & 1 \\ -1 & 1 & -1 \end{matrix} = \begin{matrix} -2 & 2 & 0 \\ 2 & -2 & 0 \\ 2 & -2 & 0 \\ -2 & 2 & 0 \end{matrix}
$$

and

$$
W^T = \begin{matrix} -2 & 2 & 2 & -2 \\ 2 & -2 & -2 & 2 \\ 0 & 0 & 0 & 0 \end{matrix}
$$

You may think that the fact that the last column of **W** is all zeros presents a problem in that when the input is X_1 and whatever the output, when this output is presented back in the backward direction, it does not produce X_1, which does not have a zero in the last component. There is a **Thresholding** function needed here. The **Thresholding** function is transparent when the activations are all either +1 or –1. It then just looks like you do an inverse mapping from bipolar values to binary, which is done by replacing each –1 by a 0. When the activation is zero, you simply leave the output of that neuron as it was in the previous cycle or iteration. We now present the **Thresholding** function for BAM outputs.

$$b_j|_{t+1} = b_j|_t \quad \begin{matrix} 1 & \text{if } y_j > 0 \\ \text{if } y_j = 0 \\ 0 & \text{if } y_j < 0 \end{matrix} \quad \text{and} \quad a_i|_{t+1} = a_i|_t \quad \begin{matrix} 1 & \text{if } x_i > 0 \\ \text{if } x_i = 0 \\ 0 & \text{if } x_i < 0 \end{matrix}$$

where x_i and y_j are the activations of neurons i and j in the input layer and output layer, respectively, and $b_j|_t$ refers to the output of the jth neuron in the output layer in the cycle t, whereas $a_i|_t$ refers to the output of the ith neuron in the input layer in the cycle t. Note that at the start, the a_i and b_j values are the same as the corresponding components in the exemplar pair being used.

If $X_1 = (1, 0, 0, 1)$ is presented to the input neurons, their activations are given by the vector $(-4, 4, 0)$. The output vector, after using the **Threshold** function just described is $(0, 1, 1)$. The last component here is supposed to be the same as the output in the previous cycle, since the corresponding activation value is 0. Since X_1 and Y_1 are one exemplar pair, the third component of Y_1 is what we need as the third component of the output in the current cycle of operation. Therefore, we fed X_1 and received Y_1. If we feed Y_1 at the other end, the activations in the A field will be $(2, -2, -2, 2)$, and the output vector will be $(1, 0, 0, 1)$, which is X_1. If you do the same with the pair X_2 and Y_2, you will see that they are heteroassociated with each other.

Let us modify our X_1 as $(1, 0, 1, 1)$. Then the weight matrix **W** becomes

$$W = \begin{matrix} 1 \\ -1 \\ 1 \\ 1 \end{matrix} \begin{bmatrix} -1 & 1 & 1 \end{bmatrix} + \begin{matrix} -1 \\ 1 \\ 1 \\ -1 \end{matrix} \begin{bmatrix} 1 & -1 & 1 \end{bmatrix} = \begin{matrix} -2 & 2 & 0 \\ 2 & -2 & 0 \\ 0 & 0 & 2 \\ -2 & 2 & 0 \end{matrix}$$

and

$$W_T = \begin{matrix} -2 & 2 & 0 & -2 \\ 2 & -2 & 0 & 2 \\ 0 & 0 & 2 & 0 \end{matrix}$$

Now this is a different set of two exemplar vector pairs. The pairs are $X_1 = (1, 0, 1, 1)$, $Y_1 = (0, 1, 1)$, and $X_2 = (0, 1, 1, 0)$, $Y_2 = (1, 0, 1)$. Naturally, the weight matrix is different, as is its transpose, correspondingly. As stated before, the weights do not change during the operation of the network with whatever inputs presented to it.

Recall

When X_1 is presented at the input layer, the activation at the output layer will give (–4, 4, 2) to which we apply the **Thresholding** function, which replaces a positive value by 1, and a negative value by 0.

This then gives us the vector (0, 1, 1) as the output, which is the same as our Y_1. Now Y_1 is passed back to the input layer through the feedback connections, and the activation of the input layer becomes the vector (2, –2, 2, 2), which after thresholding gives the output vector (1, 0, 1, 1), same as X_1. When X_2 is presented at the input layer, the activation at the output layer will give (2, –2, 2) to which the **Thresholding** function, which replaces a positive value by 1 and a negative value by 0, is applied. This then gives the vector (1, 0, 1) as the output, which is the same as Y_2. Now Y_2 is passed back to the input layer through the feedback connections to get the activation of the input layer as (–2, 2, 2, –2), which after thresholding gives the output vector (0, 1, 1, 0), which is X_2.

The two vector pairs chosen here for encoding worked out fine, and the BAM network with four neurons in field A and three neurons in field B is all set for finding a vector under heteroassociation with a given input vector.

Continuation of Example

Let us now use the vector X_3 = (1, 1, 1, 1). The vector Y_3 = (0, 1, 1) is obtained at the output layer. But the next step in which we present Y_3 in the backward direction, does not produce X_3, but gives an X_1 = (1, 0, 1, 1). We already have X_1 associated with Y_1. This means that X_3 is not associated with any vector in the output space. On the other hand, if instead of getting X_1 we obtained a different X_4 vector, and if this in the feed forward operation produced a different Y vector, then we repeat the operation of the network until no changes occur at either end. Then we will have possibly a new pair of vectors under the heteroassociation established by this BAM network.

Special Case—Complements

If a pair of (distinct) patterns X and Y are found to be heteroassociated by BAM, and if you input the complement of X, complement being obtained by interchanging the 0's and 1's in X, BAM will show that the complement of Y is the pattern associated with the complement of X. An example of this will be seen in the illustrative run of the program for C++ implementation of BAM, which follows.

C++ Implementation

In our C++ implementation of a discrete bidirectional associative memory network, we create classes for neuron and network. Other classes created are called **exemplar**, **assocpair**, **potlpair**, for the exemplar pair of vectors, associated pair of vectors, and potential pairs of vectors, respectively, for finding heteroassociation between them. We could have made one class of *pairvect*, for a pair of vectors and derived the exemplar and so on from it. The **network** class is declared as a **friend** class in these other classes mentioned. Now we present the header and source files, called *bamntwrk.h* and *bamntwrk.cpp*. Since we reused our previous code, there are a few data members of classes that we did not put to explicit use in the program. We call the **neuron** class **bmneuron** to remind us of BAM.

For our illustration run, we provided for six neurons in the input layer and five in the output layer. We used three pairs of exemplars for encoding. We used two additional input vectors, one of which is the complement of the X of an exemplar pair, after the encoding is done, to see what association will be established in these cases, or what recall will be made by the BAM network.

As expected, the complement of the Y of the exemplar is found to be associated with the complement of the X of that pair. When the second input vector is presented, however, a new pair of associated vectors is found. After the code is presented, we list the computer output also, and in bold face, as usual.

Header File

Listing 8.1   bamntwrk.h

```
//bamntwrk.h    V. Rao,   H. Rao

//Header file for BAM network program

#include <iostream.h>
#include <math.h>
#include <stdlib.h>
#define MXSIZ 10


class bmneuron
{

 protected:
    int nnbr;
    int inn,outn;
    int output;
    int activation;
    int outwt[MXSIZ];
    char *name;
    friend class network;

public:
    bmneuron() { };
    void getnrn(int,int,int,char *);
};

class exemplar
{
protected:
    int xdim,ydim;
    int v1[MXSIZ],v2[MXSIZ];
    int u1[MXSIZ],u2[MXSIZ];
    friend class network;
    friend class mtrx;

public:
    exemplar() { };
    void getexmplr(int,int,int *,int *);
    void prexmplr();
    void trnsfrm();
    void prtrnsfrm();
};

class asscpair
{
protected:
```

```
    int xdim,ydim,idn;
    int v1[MXSIZ],v2[MXSIZ];
    friend class network;

public:
    asscpair() { };
    void getasscpair(int,int,int);
    void prasscpair();
};

class potlpair
{
protected:
    int xdim,ydim;
    int v1[MXSIZ],v2[MXSIZ];
    friend class network;

public:
    potlpair() { };
    void getpotlpair(int,int);
    void prpotlpair();
};

class network
{
public:
    int   anmbr,bnmbr,flag,nexmplr,nasspr,ninpt;
    bmneuron (anrn)[MXSIZ],(bnrn)[MXSIZ];
    exemplar (e)[MXSIZ];
    asscpair (as)[MXSIZ];
    potlpair (pp)[MXSIZ];
    int outs1[MXSIZ],outs2[MXSIZ];
    int mtrx1[MXSIZ][MXSIZ],mtrx2[MXSIZ][MXSIZ];

    network() { };
    void getnwk(int,int,int,int [][6],int [][5]);
    void compr1(int,int);
    void compr2(int,int);
    void prwts();
    void iterate();
    void findassc(int *);
    void asgninpt(int *);
    void asgnvect(int,int *,int *);
    void comput1();
    void comput2();
    void prstatus();
};
```

Source File

Listing 8.2   bamntwrk.cpp

```
//bamntwrk.cpp    V. Rao, H. Rao

//Source file for BAM network program

#include "bamntwrk.h"

void bmneuron::getnrn(int m1,int m2,int m3,char *y)
{
int i;
name = y;
nnbr = m1;
outn = m2;
inn  = m3;

for(i=0;i<outn;++i){
    outwt[i] = 0 ;
    }

output = 0;
activation = 0;
}

void exemplar::getexmplr(int k,int l,int *b1,int *b2)
{
int i2;
xdim = k;
ydim = l;

for(i2=0;i2<xdim;++i2){
    v1[i2] = b1[i2]; }

for(i2=0;i2<ydim;++i2){
    v2[i2] = b2[i2]; }
}

void exemplar::prexmplr()
{
int i;
cout<<"\nX vector you gave is:\n";

for(i=0;i<xdim;++i){
    cout<<v1[i]<<"  ";}

cout<<"\nY vector you gave is:\n";

for(i=0;i<ydim;++i){
    cout<<v2[i]<<"  ";}
```

```
cout<<"\n";
}

void exemplar::trnsfrm()
{
int i;

for(i=0;i<xdim;++i){
    u1[i] = 2*v1[i] -1;}

for(i=0;i<ydim;++i){
    u2[i] = 2*v2[i] - 1;}

}

void exemplar::prtrnsfrm()
{
int i;
cout<<"\nbipolar version of X vector you gave is:\n";

for(i=0;i<xdim;++i){
    cout<<u1[i]<<"  ";}

cout<<"\nbipolar version of Y vector you gave is:\n";

for(i=0;i<ydim;++i){
    cout<<u2[i]<<"  ";}

cout<<"\n";
}

void asscpair::getasscpair(int i,int j,int k)
{
idn = i;
xdim = j;
ydim = k;
}

void asscpair::prasscpair()
{
int i;
cout<<"\nX vector in the associated pair no. "<<idn<<"
    is:\n";

for(i=0;i<xdim;++i){
    cout<<v1[i]<<"  ";}

cout<<"\nY vector in the associated pair no. "<<idn<<"
    is:\n";
```

```
for(i=0;i<ydim;++i){
    cout<<v2[i]<<"  ";}

cout<<"\n";
}

void potlpair::getpotlpair(int k,int j)
{

xdim = k;
ydim = j;

}

void potlpair::prpotlpair()
{
int i;
cout<<"\nX vector in possible associated pair is:\n";

for(i=0;i<xdim;++i){
    cout<<v1[i]<<"  ";}

cout<<"\nY vector in possible associated pair is:\n";

for(i=0;i<ydim;++i){
    cout<<v2[i]<<"  ";}

cout<<"\n";
}

void network::getnwk(int k,int l,int k1,int b1[][6],int
    b2[][5])
{
anmbr = k;
bnmbr = l;
nexmplr = k1;
nasspr = 0;
ninpt = 0;
int i,j,i2;
flag =0;
char *y1="ANEURON", *y2="BNEURON" ;

for(i=0;i<nexmplr;++i){
    e[i].getexmplr(anmbr,bnmbr,b1[i],b2[i]);
    e[i].prexmplr();
    e[i].trnsfrm();
    e[i].prtrnsfrm();
    }
```

```
for(i=0;i<anmbr;++i){
    anrn[i].bmneuron::getnrn(i,bnmbr,0,y1);}

for(i=0;i<bnmbr;++i){
    bnrn[i].bmneuron::getnrn(i,0,anmbr,y2);}

for(i=0;i<anmbr;++i){

    for(j=0;j<bnmbr;++j){
        mtrx1[i][j]  = 0;

        for(i2=0;i2<nexmplr;++i2){
            mtrx1[i][j]  += e[i2].u1[i]*e[i2].u2[j];}

        mtrx2[j][i] = mtrx1[i][j];
        anrn[i].outwt[j] = mtrx1[i][j];
        bnrn[j].outwt[i] = mtrx2[j][i];
    }
}

prwts();
cout<<"\n";
}

void network::asgninpt(int *b)
{
int i;
cout<<"\n";

for(i=0;i<anmbr;++i){
    anrn[i].output = b[i];
    outs1[i] = b[i];
    }

}

void network::compr1(int j,int k)
{
int i;

for(i=0;i<anmbr;++i){

    if(pp[j].v1[i] != pp[k].v1[i]) flag = 1;
    break;
    }

}

void network::compr2(int j,int k)
{
int i;
```

```
for(i=0;i<anmbr;++i){

    if(pp[j].v2[i] != pp[k].v2[i]) flag = 1;
    break;}

}

void network::comput1()
{
int j;

for(j=0;j<bnmbr;++j){
    int ii1;
    int c1 =0,d1;
    cout<<"\n";

    for(ii1=0;ii1<anmbr;++ii1){
        d1 = outs1[ii1] * mtrx1[ii1][j];
        c1 += d1;
        }

    bnrn[j].activation = c1;
    cout<<"\n output layer neuron  "<<j<<" activation is
        "<<c1<<"\n";

    if(bnrn[j].activation <0) {

        bnrn[j].output = 0;
        outs2[j] = 0;}

    else

        if(bnrn[j].activation>0) {

            bnrn[j].output = 1;
            outs2[j] = 1;}

            else

            {cout<<"\n A 0 is obtained, use previous output
                value \n";

            if(ninpt<=nexmplr){

                bnrn[j].output = e[ninpt-1].v2[j];}

            else

                { bnrn[j].output = pp[0].v2[j];}
```

```
                outs2[j] = bnrn[j].output; }

        cout<<"\n output layer neuron  "<<j<<" output is
            "<<bnrn[j].output<<"\n";
        }
}


void network::comput2()
{
int i;

for(i=0;i<anmbr;++i){
    int ii1;
    int c1=0;

    for(ii1=0;ii1<bnmbr;++ii1){
        c1 += outs2[ii1] * mtrx2[ii1][i];  }

    anrn[i].activation = c1;
    cout<<"\ninput layer neuron "<<i<<"activation is \
        "<<c1<<"\n";

    if(anrn[i].activation <0 ){

        anrn[i].output = 0;
        outs1[i] = 0;}

    else

        if(anrn[i].activation >0 ) {

            anrn[i].output = 1;
            outs1[i] = 1;
            }

        else

        { cout<<"\n A 0 is obtained, use previous value if \
            available\n";

        if(ninpt<=nexmplr){

            anrn[i].output = e[ninpt-1].v1[i];}

        else

            {anrn[i].output = pp[0].v1[i];}

        outs1[i] = anrn[i].output;}
```

```
            cout<<"\n input layer neuron  "<<i<<" output is \
                "<<anrn[i].output<<"\n";
            }
    }

    void network::asgnvect(int j1,int *b1,int *b2)
    {
    int  j2;

    for(j2=0;j2<j1;++j2){
        b2[j2] = b1[j2];}

    }

    void network::prwts()
    {
    int i3,i4;
    cout<<"\n weights--  input layer to output layer: \n\n";

    for(i3=0;i3<anmbr;++i3){

        for(i4=0;i4<bnmbr;++i4){
            cout<<anrn[i3].outwt[i4]<<"  ";}

        cout<<"\n"; }

    cout<<"\n";

    cout<<"\nweights--  output layer to input layer: \n\n";

    for(i3=0;i3<bnmbr;++i3){

        for(i4=0;i4<anmbr;++i4){
            cout<<bnrn[i3].outwt[i4]<<"  ";}

        cout<<"\n";   }

    cout<<"\n";
    }

    void network::iterate()
    {
    int i1;

    for(i1=0;i1<nexmplr;++i1){
        findassc(e[i1].v1);
        }

    }

    void network::findassc(int *b)
```

```
{
int j;
flag = 0;
    asgninpt(b);
    ninpt ++;
    cout<<"\nInput vector is:\n" ;

    for(j=0;j<6;++j){
        cout<<b[j]<<" ";}

    cout<<"\n";
    pp[0].getpotlpair(anmbr,bnmbr);
    asgnvect(anmbr,outs1,pp[0].v1);

    comput1();

    if(flag>=0){
        asgnvect(bnmbr,outs2,pp[0].v2);

        cout<<"\n";
        pp[0].prpotlpair();
        cout<<"\n";

        comput2(); }

    for(j=1;j<MXSIZ;++j){
        pp[j].getpotlpair(anmbr,bnmbr);
        asgnvect(anmbr,outs1,pp[j].v1);

        comput1();

        asgnvect(bnmbr,outs2,pp[j].v2);

        pp[j].prpotlpair();
        cout<<"\n";

        compr1(j,j-1);
        compr2(j,j-1);

        if(flag == 0) {

            int j2;
            nasspr += 1;
            j2 = nasspr;

            as[j2].getasscpair(j2,anmbr,bnmbr);
            asgnvect(anmbr,pp[j].v1,as[j2].v1);
            asgnvect(bnmbr,pp[j].v2,as[j2].v2);

            cout<<"\nPATTERNS ASSOCIATED:\n";
            as[j2].prasscpair();
```

```
                j = MXSIZ ;
            }

        else

            if(flag == 1)

                {
                flag = 0;
                comput1();
                }

    }
}

void network::prstatus()
{
int j;
cout<<"\nTHE FOLLOWING ASSOCIATED PAIRS WERE FOUND BY \
    BAM\n\n";

for(j=1;j<=nasspr;++j){
    as[j].prasscpair();
    cout<<"\n";}

}

void main()
{
int ar = 6, br = 5, nex = 3;
int
inptv[][6]={1,0,1,0,1,0,1,1,1,0,0,0,0,1,1,0,0,0,0,1,0,1,0,1,
    1,1,1,1,1,1};
int outv[][5]={1,1,0,0,1,0,1,0,1,1,1,0,0,1,0};

cout<<"\n\nTHIS PROGRAM IS FOR A BIDIRECTIONAL ASSOCIATIVE \
    MEMORY NETWORK.\n";
cout<<" THE NETWORK ISSET UP FOR ILLUSTRATION WITH "<<ar<<" \
    INPUT NEURONS, AND "<<br;
cout<<" OUTPUT NEURONS.\n"<<nex<<" exemplars are used to \
    encode \n";

static network bamn;
bamn.getnwk(ar,br,nex,inptv,outv) ;
bamn.iterate();
bamn.findassc(inptv[3]);
bamn.findassc(inptv[4]);
bamn.prstatus();

}
```

Program Output

The output from an illustrative run of the program is listed next. The computer output is in bold face. You will notice that we provided for a lot of information to be output as the program is executed. The fourth vector we input is not from an exemplar, but it is the complement of the X of the first exemplar pair. The network found the complement of the Y of this exemplar to be associated with this input. The fifth vector we input is (1, 1, 1, 1, 1, 1). But BAM recalled in this case the complement pair for the third exemplar pair, which is **X** = (1, 0, 0, 1, 1, 1) and **Y** = (0, 1, 1, 0, 1). You notice that the Hamming distance of this input pattern from the X's of the three exemplars are 3, 3, and 4, respectively. The hamming distance from the complement of the X of the first exemplar pair is also 3. But the Hamming distance of (1, 1, 1, 1) from the X of the complement of the X of the third exemplar pair is only 2. It would be instructive if you would use the input pattern (1, 1, 1, 1, 1, 0) to see to what associated pair it leads. Now the output.

THIS PROGRAM IS FOR A BIDIRECTIONAL ASSOCIATIVE
MEMORY NETWORK.
THE NETWORK IS SET UP FOR ILLUSTRATION WITH SIX INPUT
NEURONS AND FIVE OUTPUT NEURONS.
Three exemplars are used to encode

X vector you gave is:
1 0 1 0 1 0
Y vector you gave is:
1 1 0 0 1

bipolar version of X vector you gave is:
1 -1 1 -1 1 -1
bipolar version of Y vector you gave is:
1 1 -1 -1 1

X vector you gave is:
1 1 1 0 0 0
Y vector you gave is:
0 1 0 1 1

bipolar version of X vector you gave is:
1 1 1 -1 -1 -1
bipolar version of Y vector you gave is:
-1 1 -1 1 1

X vector you gave is:
0 1 1 0 0 0

Y vector you gave is:
1 0 0 1 0

bipolar version of X vector you gave is:
-1 1 1 -1 -1 -1
bipolar version of Y vector you gave is:
1 -1 -1 1 -1

   weights-- input layer to output layer:

   -1   3 -1 -1   3
   -1   1 -1   3 -1
    1   1 -3   1   1
   -1 -1   3 -1 -1
    1   1   1 -3   1
   -1 -1   3 -1 -1

   weights-- output layer to input layer:

   -1 -1   1 -1   1 -1
    3 -1   1 -1   1 -1
   -1 -1 -3   3   1   3
   -1   3   1 -1 -3 -1
    3 -1   1 -1   1 -1

Input vector is:
1 0 1 0 1 0

   output layer neuron 0 activation is 1

   output layer neuron 0 output is 1

   output layer neuron 1 activation is 5

   output layer neuron 1 output is 1

   output layer neuron 2 activation is -3

   output layer neuron 2 output is 0

   output layer neuron 3 activation is -3

   output layer neuron 3 output is 0

output layer neuron 4 activation is 5

output layer neuron 4 output is 1

X vector in possible associated pair is:
1 0 1 0 1 0
Y vector in possible associated pair is:
1 1 0 0 1

input layer neuron 0 activation is 5

 input layer neuron 0 output is 1

input layer neuron 1 activation is -3

 input layer neuron 1 output is 0

input layer neuron 2 activation is 3

 input layer neuron 2 output is 1

input layer neuron 3 activation is -3

 input layer neuron 3 output is 0

input layer neuron 4 activation is 3

 input layer neuron 4 output is 1

input layer neuron 5 activation is -3

 input layer neuron 5 output is 0

 output layer neuron 0 activation is 1

 output layer neuron 0 output is 1

 output layer neuron 1 activation is 5

 output layer neuron 1 output is 1

 output layer neuron 2 activation is -3

 output layer neuron 2 output is 0

output layer neuron 3 activation is -3

output layer neuron 3 output is 0

output layer neuron 4 activation is 5

output layer neuron 4 output is 1

X vector in possible associated pair is:
1 0 1 0 1 0
Y vector in possible associated pair is:
1 1 0 0 1

PATTERNS ASSOCIATED:

X vector in the associated pair no. 1 is:
1 0 1 0 1 0
Y vector in the associated pair no. 1 is:
1 1 0 0 1

Input vector is:
1 1 1 0 0 0

```
// We get here more of the detailed output as in the previous
case. We will simply not present it here.
```

PATTERNS ASSOCIATED:

X vector in the associated pair no. 1 is:
1 1 1 0 0 0
Y vector in the associated pair no. 1 is:
0 1 0 1 1

Input vector is:
0 1 1 0 0 0

output layer neuron 0 activation is 0

A 0 is obtained, use previous output value

output layer neuron 0 output is 1

output layer neuron 1 activation is 0

A 0 is obtained, use previous output value

output layer neuron 1 output is 0

output layer neuron 2 activation is -4

output layer neuron 2 output is 0

output layer neuron 3 activation is 4

output layer neuron 3 output is 1

output layer neuron 4 activation is 0

A 0 is obtained, use previous output value

output layer neuron 4 output is 0

X vector in possible associated pair is:
0 1 1 0 0 0
Y vector in possible associated pair is:
1 0 0 1 0

// We get here more of the detailed output as in the previous
case. We will simply not present it here.

PATTERNS ASSOCIATED:

X vector in the associated pair no. 1 is:
0 1 1 0 0 0
Y vector in the associated pair no. 1 is:
1 0 0 1 0

Input vector is:
0 1 0 1 0 1

// We get here more of the detailed output as in the previous
case. We will simply not present it here.

X vector in possible associated pair is:
0 1 0 1 0 1
Y vector in possible associated pair is:
0 0 1 1 0

```
// We get here more of the detailed output as in the previous
case. We will simply not present it here.
```

X vector in possible associated pair is:
0 1 0 1 0 1
Y vector in possible associated pair is:
0 0 1 1 0

PATTERNS ASSOCIATED:

X vector in the associated pair no. 1 is:
0 1 0 1 0 1
Y vector in the associated pair no. 1 is:
0 0 1 1 0

Input vector is:
1 1 1 1 1 1

output layer neuron 0 activation is -2

output layer neuron 0 output is 0

output layer neuron 1 activation is 2

output layer neuron 1 output is 1

output layer neuron 2 activation is 2

output layer neuron 2 output is 1

output layer neuron 3 activation is -2

output layer neuron 3 output is 0

output layer neuron 4 activation is 2

output layer neuron 4 output is 1

X vector in possible associated pair is:
1 1 1 1 1 1
Y vector in possible associated pair is:
0 1 1 0 1

input layer neuron 0 activation is 5

input layer neuron 0 output is 1

input layer neuron 1 activation is -3

input layer neuron 1 output is 0

input layer neuron 2 activation is -1

input layer neuron 2 output is 0

input layer neuron 3 activation is 1

input layer neuron 3 output is 1

input layer neuron 4 activation is 3

input layer neuron 4 output is 1

input layer neuron 5 activation is 1

input layer neuron 5 output is 1

output layer neuron 0 activation is -2

output layer neuron 0 output is 0

output layer neuron 1 activation is 2

output layer neuron 1 output is 1

output layer neuron 2 activation is 6

output layer neuron 2 output is 1

output layer neuron 3 activation is -6

output layer neuron 3 output is 0

output layer neuron 4 activation is 2

output layer neuron 4 output is 1

X vector in possible associated pair is:

1 0 0 1 1 1
Y vector in possible associated pair is:
0 1 1 0 1

PATTERNS ASSOCIATED:

X vector in the associated pair no. 1 is:
1 0 0 1 1 1
Y vector in the associated pair no. 1 is:
0 1 1 0 1

THE FOLLOWING ASSOCIATED PAIRS WERE FOUND BY BAM

X vector in the associated pair no. 1 is: `//first exemplar pair`
1 0 1 0 1 0
Y vector in the associated pair no. 1 is:
1 1 0 0 1

X vector in the associated pair no. 2 is: `//second exemplar pair`
1 1 1 0 0 0
Y vector in the associated pair no. 2 is:
0 1 0 1 1

X vector in the associated pair no. 3 is: `//third exemplar pair`
0 1 1 0 0 0
Y vector in the associated pair no. 3 is:
1 0 0 1 0

X vector in the associated pair no. 4 is: `//complement of X of the`
0 1 0 1 0 1   `first exemplar pair`

Y vector in the associated pair no. 4 is: `//complement of Y of the`
0 0 1 1 0   `first exemplar pair`

X vector in the associated pair no. 5 is: `//input was X = (1, 1, 1,`
1 0 0 1 1 1   `1, 1) but result was comple-`
                                                 `ment of third exemplar pair`
Y vector in the associated pair no. 5 is: `with X of which Hamming`
0 1 1 0 1   `distance is the least.`

Summary

In this chapter bidirectional associative memories are presented. The development of these memories is largely due to Kosko. They share with Adaptive Resonance Theory the feature of resonance between the two layers in the network. The bidirectional associative memories (BAM) network finds heteroassociation between binary patterns, and these are converted to bipolar values to determine the connection weight matrix. Even though there are connections in both directions between neurons in the two layers, essentially only one weight matrix is involved. You use the transpose of this weight matrix for the connections in the opposite direction. When one input at one end leads to some output at the other end, which in turn leads to output that is the same as the previous input, resonance is reached and an associated pair is found.

Chapter
9

FAM: Fuzzy Associative Memory

Introduction

ere we will be concerned with fuzzy sets and their elements, both for input and output of a neural network. Every element of a fuzzy set has a degree of membership in the set. Unless this degree of membership is 1, an element does not belong to the set (in the sense of elements of an ordinary set belonging to the set). If there is a 50% chance of rain tomorrow, then tomorrow belongs to the set of rainy days with a membership degree of 0.5. In a neural network of a fuzzy system the inputs, the outputs, and the connection weights all belong fuzzily to the spaces that define them. The weight matrix will be a fuzzy matrix, and the activations of the neurons in such a network have to be determined by rules of fuzzy logic and fuzzy set operations.

As you may know, an expert system uses what are called *crisp rules* and applies them sequentially. The advantage in casting the same problem in a fuzzy system is that the rules you work with do not have to be crisp, and the processing is done in parallel. What the fuzzy systems can determine is a fuzzy association. These associations can be modified, and the underlying phenomena better understood, as experience is gained. That is one of the reasons for their growing popularity in applications. When we try to relate two things through a process of trial and error, we will be implicitly and intuitively establishing an association that is gradually modified and perhaps bettered in some sense. Several fuzzy variables may be present in such an exercise, and we seem to process in parallel instinctively. That we did not have full knowledge at the beginning is not a hindrance; there is some difference in using probabilities and using fuzzy logic as well. The degree of membership assigned for an element of a set does not have to be as firm as the assignment of a probability.

The degree of membership is, like a probability, a real number between 0 and 1. The closer it is to 1, the less ambiguous is the membership of the element in the set concerned. Suppose you have a set that may or may not contain three elements, say, a, b, and c. Then the fuzzy set representation of it would be by the ordered triple (m_a, m_b, m_c), which is called the *fit vector*, and its components are called *fit values*. For example, the triple (0.5, 0.5, 0.5) shows that each of a, b, and c, have a membership equal to only one-half. This triple itself will describe the fuzzy set. It can also be thought of as a point in the three-dimensional space. None of such points will be outside the unit cube. When the number of elements is higher, the corresponding points would be on or inside the unit hypercube.

It is interesting to note that this fuzzy set, given by the triple (0.5, 0.5, 0.5) is its own complement, something that does not happen with regular sets. The complement is the set that shows the degrees of nonmembership.

The *height* of a fuzzy set is the maximum of its fit values, and the fuzzy set is said to be *normal*, if its height is 1. The fuzzy set with fit vector (0.3, 0.7, 0.4) has height 0.7, and it is not a normal fuzzy set. However, by introducing an additional dummy component with fit value 1, we can extend it into a normal fuzzy set. The desirability of normalcy of a fuzzy set will become apparent when we talk about recall in fuzzy associative memories.

The subset relationship is also different for fuzzy sets from the way it is defined for regular sets. For example, if you have a fuzzy set given by the triple (0.3, 0.7, 0.4) , then any fuzzy set with a triple (a, b, c) such that a ≤ .3, b ≤ .7, and c ≤ .4, is its fuzzy subset. For example, the fuzzy set given by the triple (0.1, 0.7, 0) is a subset of the fuzzy set (0.3, 0.7, 0.4).

Association

Consider two fuzzy sets, one perhaps referring to the popularity of (or interest in) an exhibition and the other to the price of admission. The popularity could be very high, high, fair, low, or very low. Accordingly, the fit vector will have five components. The price of admission could be high, modest, or low. Its fit vector has three components. A fuzzy associative memory system then has the association (popularity, price), and the fuzzy set pair encodes this association. You need a different fuzzy associative memory system for each fuzzy set pair and the corresponding association.

We describe the encoding process and the recall in the following sections. Once encoding is completed, associations between subsets of the first fuzzy set and subsets of the second fuzzy set are also established by the same fuzzy associative memory system.

FAM Neural Network

The neural network for a fuzzy associative memory has an input and an output layer, with full connections in both directions, just as in a BAM neural network. Figure 9.1 shows the layout. To avoid cluttering, the figure shows the network with three neurons in field A and two neurons in field B, and only some of the connection weights. The general case is analogous.

Encoding

Encoding for fuzzy associative memory systems is similar in concept to the encoding process for bidirectional associative memories, with some differences in the operations involved. In the BAM encoding, bipolar versions of binary vectors are used just for encoding. Matrices of the type $X_i^T Y_i$ are added together to get the connection weight matrix. There are two basic operations with the elements of these vectors. They are multiplication and addition of products. There is no conversion of the fit values before the encoding by the fuzzy sets. The multiplication of elements is replaced by the operation of taking the **minimum** and addition is replaced by the operation of taking the **maximum**.

There are two methods for encoding. The method just described is what is called *max–min composition*. It is used to get the connection weight matrix and also to get the outputs of neurons in the fuzzy associative memory neural network. The second method is called *correlation–product encoding*. It is obtained the same way as a BAM connection weight matrix is obtained. Max–min composition is what is frequently used in practice, and we will confine our attention to this method.

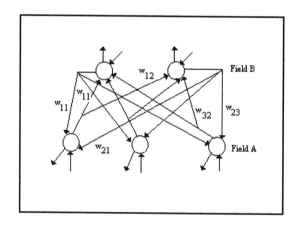

FIGURE 9.1 *Fuzzy Associative Memory Neural Network.*

Example of Encoding

Suppose the two fuzzy sets we use to encode have the fit vectors (0.3, 0.7, 0.4, 0.2) and (0.4, 0.3, 0.9). Then the matrix W is obtained by using max–min composition as follows.

```
        0.3 [0.4 0.3 0.9]    min(0.3,0.4) min(0.3,0.3) min(0.3,0.9)    0.3 0.3 0.3
W =     0.7                = min(0.7,0.4) min(0.7,0.3) min(0.7,0.9) = 0.4 0.3 0.7
        0.4                  min(0.4,0.4) min(0.4,0.3) min(0.4,0.9)    0.4 0.3 0.4
        0.2                  min(0.2,0.4) min(0.2,0.3) min(0.2,0.9)    0.2 0.2 0.2
```

Recall for the Example

If we input the fit vector (0.3, 0.7, 0.4, 0.2), the output (b_1, b_2, b_3) is determined as follows, using b_j = max(min(a_1, w_{1j}), ..., min(a_m, w_{mj}), where m is the dimension of the 'a' fit vector, and w_{ij} is the ith row, jth column element of the matrix W.

```
b₁ = max(min(0.3, 0.3), min(0.7, 0.4), min(0.4, 0.4),
     min(0.2, 0.2)) =  max(0.3, 0.4, 0.4, 0.2) = 0.4
b₂ = max(min(0.3, 0.3), min(0.7, 0.3), min(0.4, 0.3),
     min(0.2, 0.2)) = max( 0.3, 0.3, 0.3, 0.2 ) = 0.3
b₃ = max(min(0.3, 0.3), min(0.7, 0.7), min(0.4, 0.4),
     min(0.2, 0.2)) = max (0.3, 0.7, 0.4, 0.2) = 0.7
```

The output vector (0.4, 0.3, 0.7) is not the same as the second fit vector used, namely (0.4, 0.3, 0.9), but it is a subset of it, so the recall is not perfect. If you input the vector (0.4, 0.3, 0.7) in the opposite direction, using the transpose of the matrix W, the output is (0.3, 0.7, 0.4, 0.2), showing resonance. If on the other hand you input (0.4, 0.3, 0.9) at that end, the output vector is (0.3, 0.7, 0.4, 0.2), which in turn causes in the other direction an output of (0.4, 0.3, 0.7) at which time there is resonance. Can we foresee these results? The following will explain.

Recall

Let us use the operator **o** to denote max–min composition. The theorem that tells you when perfect recall occurs when the weight matrix is obtained using the max–min composition of fit vectors U and V says,

(i) U o W = V if and only if height (U) ≥ height (V).

(ii) V o W^T = U if and only if height (V) ≥ height (U).

It also says that if X and Y are arbitrary fit vectors with the same dimensions as U and V, then

(iii) X o W ⊂ V.

(iv) Y o W^T ⊂ U.

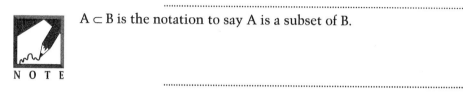

A ⊂ B is the notation to say A is a subset of B.

N O T E

In the previous example, height of (0.3, 0.7, 0.4, 0.2) is 0.7, and height of (0.4, 0.3, 0.9) is 0.9. Therefore (0.4, 0.3, 0.9) as input produced (0.3, 0.7, 0.4, 0.2) as output, but (0.3, 0.7, 0.4, 0.2) as input produced only a subset of (0.4, 0.3, 0.9). That both (0.4, 0.3, 0.7) and (0.4, 0.3, 0.9) gave the same output, (0.3, 0.7, 0.4, 0.2) is in accordance with the corollary to the above theorem, which states that if (X, Y) is a fuzzy associated memory, and if X is a subset of X', then (X', Y) is also a fuzzy associated memory.

C++ Implementation

We use the classes we created for BAM implementation in C++, except that we call the neuron class **fzneuron**, and we do not need some of the methods or functions in the **network** class. The header file, the source file, and the output from an illustrative run of the program are given in the following. The header file is called **fuzzyam.hpp**, and the source file is called **fuzzyam.cpp**. The computer output is in bold face as before.

Header File

Listing 9.1   fuzzyam.h
```
//fuzzyam.h   V. Rao, H. Rao

#include <iostream.h>
#define MXSIZ 10

class fzneuron
{
```

```
protected:
    int nnbr;
    int inn,outn;
    float output;
    float activation;
    float outwt[MXSIZ];
    char *name;
    friend class network;

public:
    fzneuron() { };
    void getnrn(int,int,int,char *);
};

class exemplar
{
protected:
    int xdim,ydim;
    float v1[MXSIZ],v2[MXSIZ];
    friend class network;

public:
    exemplar() { };
    void getexmplr(int,int,float *,float *);
    void prexmplr();
};

class asscpair
{
protected:
    int xdim,ydim,idn;
    float v1[MXSIZ],v2[MXSIZ];
    friend class network;

public:
    asscpair() { };
    void getasscpair(int,int,int);
    void prasscpair();
};

class potlpair
{
protected:
    int xdim,ydim;
    float v1[MXSIZ],v2[MXSIZ];
    friend class network;

public:
    potlpair() { };
    void getpotlpair(int,int);
    void prpotlpair();
```

```
};

class network
{
public:
    int   anmbr,bnmbr,flag,nexmplr,nasspr,ninpt;
    fzneuron (anrn)[MXSIZ],(bnrn)[MXSIZ];
    exemplar (e)[MXSIZ];
    asscpair (as)[MXSIZ];
    potlpair (pp)[MXSIZ];
    float outs1[MXSIZ],outs2[MXSIZ];
    double mtrx1[MXSIZ][MXSIZ],mtrx2[MXSIZ][MXSIZ];

    network() { };
    void getnwk(int,int,int,float [][6],float [][4]);
    void compr1(int,int);
    void compr2(int,int);
    void prwts();
    void iterate();
    void findassc(float *);
    void asgninpt(float *);
    void asgnvect(int,float *,float *);
    void comput1();
    void comput2();
    void prstatus();
};
```

Source File

Listing 9.2   fuzzyam.cpp
```
//fuzzyam.cpp    V. Rao, H. Rao

#include "fuzzyam.h"

float max(float x,float y)
{
float u;
u = ((x>y) ? x : y );
return u;
}

float min(float x,float y)
{
float u;
u =( (x>y) ? y : x) ;
return u;
}
```

```
void fzneuron::getnrn(int m1,int m2,int m3,char *y)
{
int i;
name = y;
nnbr = m1;
outn = m2;
inn  = m3;

for(i=0;i<outn;++i){
    outwt[i] = 0 ;
    }

output = 0;
activation = 0;
}

void exemplar::getexmplr(int k,int l,float *b1,float *b2)
{
int i2;
xdim = k;
ydim = l;

for(i2=0;i2<xdim;++i2){
    v1[i2] = b1[i2]; }

for(i2=0;i2<ydim;++i2){
    v2[i2] = b2[i2]; }

}

void exemplar::prexmplr()
{
int i;
cout<<"\nX vector you gave is:\n";

for(i=0;i<xdim;++i){
    cout<<v1[i]<<"   ";}

cout<<"\nY vector you gave is:\n";

for(i=0;i<ydim;++i){
    cout<<v2[i]<<"   ";}

cout<<"\n";
}

void asscpair::getasscpair(int i,int j,int k)
{
idn = i;
xdim = j;
```

```
ydim = k;
}

void asscpair::prasscpair()
{
int i;
cout<<"\nX vector in the associated pair no. "<<idn<<" \
    is:\n";

for(i=0;i<xdim;++i){
    cout<<v1[i]<<"   ";}

cout<<"\nY vector in the associated pair no. "<<idn<<" \
    is:\n";

for(i=0;i<ydim;++i){
    cout<<v2[i]<<"   ";}

cout<<"\n";
}

void potlpair::getpotlpair(int k,int j)
{
xdim = k;
ydim = j;

}

void potlpair::prpotlpair()
{
int i;
cout<<"\nX vector in possible associated pair is:\n";

for(i=0;i<xdim;++i){
    cout<<v1[i]<<"   ";}

cout<<"\nY vector in possible associated pair is:\n";

for(i=0;i<ydim;++i){
    cout<<v2[i]<<"   ";}

cout<<"\n";
}

void network::getnwk(int k,int l,int k1,float b1[][6],float
    b2[][4])
{
anmbr = k;
bnmbr = l;
nexmplr = k1;
nasspr = 0;
```

```
ninpt = 0;
int i,j,i2;
float tmp1,tmp2;
flag =0;
char *y1="ANEURON", *y2="BNEURON" ;

for(i=0;i<nexmplr;++i){
    e[i].getexmplr(anmbr,bnmbr,b1[i],b2[i]);
    e[i].prexmplr();
    cout<<"\n";
    }

for(i=0;i<anmbr;++i){
    anrn[i].fzneuron::getnrn(i,bnmbr,0,y1);}

for(i=0;i<bnmbr;++i){
    bnrn[i].fzneuron::getnrn(i,0,anmbr,y2);}

for(i=0;i<anmbr;++i){

    for(j=0;j<bnmbr;++j){
        tmp1 = 0.0;

        for(i2=0;i2<nexmplr;++i2){
            tmp2 = min(e[i2].v1[i],e[i2].v2[j]);
            tmp1 = max(tmp1,tmp2);
        }

        mtrx1[i][j] = tmp1;
        mtrx2[j][i] = mtrx1[i][j];
        anrn[i].outwt[j] = mtrx1[i][j];
        bnrn[j].outwt[i] = mtrx2[j][i];
        }

    }

prwts();
cout<<"\n";
}

void network::asgninpt(float *b)
{
int i,j;
cout<<"\n";

for(i=0;i<anmbr;++i){
    anrn[i].output = b[i];
    outs1[i] = b[i];
    }

}
```

```cpp
void network::compr1(int j,int k)
{
int i;

for(i=0;i<anmbr;++i){

    if(pp[j].v1[i] != pp[k].v1[i]) flag = 1;

    break;
    }
}

void network::compr2(int j,int k)
{
int i;

for(i=0;i<anmbr;++i){

    if(pp[j].v2[i] != pp[k].v2[i]) flag = 1;

    break;}

}

void network::comput1()
{
int j;

for(j=0;j<bnmbr;++j){
    int ii1;
    float c1 =0.0,d1;
    cout<<"\n";

    for(ii1=0;ii1<anmbr;++ii1){
        d1 = min(outs1[ii1],mtrx1[ii1][j]);
        c1 = max(c1,d1);
        }

    bnrn[j].activation = c1;
    cout<<"\n output layer neuron  "<<j<<" activation is \
        "<<c1<<"\n";
    bnrn[j].output = bnrn[j].activation;
    outs2[j] = bnrn[j].output;
    cout<<"\n output layer neuron  "<<j<<" output is \
        "<<bnrn[j].output<<"\n";
    }
}

void network::comput2()
{
```

```
int i;

for(i=0;i<anmbr;++i){
    int ii1;
    float c1=0.0,d1;

    for(ii1=0;ii1<bnmbr;++ii1){
        d1 = min(outs2[ii1],mtrx2[ii1][i]);
        c1 = max(c1,d1);}

    anrn[i].activation = c1;
    cout<<"\ninput layer neuron "<<i<<"activation is \
        "<<c1<<"\n";
    anrn[i].output = anrn[i].activation;
    outs1[i] = anrn[i].output;
    cout<<"\n input layer neuron  "<<i<<"output is \
        "<<anrn[i].output<<"\n";
    }

}

void network::asgnvect(int j1,float *b1,float *b2)
{
int  j2;

for(j2=0;j2<j1;++j2){
    b2[j2] = b1[j2];}

}

void network::prwts()
{
int i3,i4;
cout<<"\n  weights--  input layer to output layer: \n\n";

for(i3=0;i3<anmbr;++i3){

    for(i4=0;i4<bnmbr;++i4){
        cout<<anrn[i3].outwt[i4]<<"  ";}

    cout<<"\n"; }

cout<<"\n";

cout<<"\nweights--  output layer to input layer: \n\n";

for(i3=0;i3<bnmbr;++i3){

    for(i4=0;i4<anmbr;++i4){
        cout<<bnrn[i3].outwt[i4]<<"  ";}
```

```
        cout<<"\n";   }

   cout<<"\n";
   }

   void network::iterate()
   {
   int i1;

   for(i1=0;i1<nexmplr;++i1){
       findassc(e[i1].v1);
       }

   }

   void network::findassc(float *b)
   {
   int j;
   flag = 0;
   asgninpt(b);
       ninpt ++;
       cout<<"\nInput vector is:\n" ;

       for(j=0;j<6;++j){
           cout<<b[j]<<" ";};

       cout<<"\n";
       pp[0].getpotlpair(anmbr,bnmbr);

   asgnvect(anmbr,outs1,pp[0].v1);
   comput1();

   if(flag>=0){

       asgnvect(bnmbr,outs2,pp[0].v2);

       cout<<"\n";
       pp[0].prpotlpair();
       cout<<"\n";

       comput2(); }

   for(j=1;j<MXSIZ;++j){
       pp[j].getpotlpair(anmbr,bnmbr);
       asgnvect(anmbr,outs1,pp[j].v1);
       comput1();

       asgnvect(bnmbr,outs2,pp[j].v2);

       pp[j].prpotlpair();
       cout<<"\n";
```

```
        compr1(j,j-1);
        compr2(j,j-1);

        if(flag == 0) {

            int j2;
            nasspr += 1;
            j2 = nasspr;
            as[j2].getasscpair(j2,anmbr,bnmbr);
            asgnvect(anmbr,pp[j].v1,as[j2].v1);
            asgnvect(bnmbr,pp[j].v2,as[j2].v2);

            cout<<"\nPATTERNS ASSOCIATED:\n";
            as[j2].prasscpair();
            j = MXSIZ ;
            }

        else

            if(flag == 1)
                {
                flag = 0;
                comput1();
                }

            }

    }

void network::prstatus()
{
int j;
cout<<"\nTHE FOLLOWING ASSOCIATED PAIRS WERE FOUND BY FUZZY
    AM\n\n";

for(j=1;j<=nasspr;++j){
    as[j].prasscpair();
    cout<<"\n";}

}

void main()
{
int ar = 6, br = 4, nex = 1;
float
inptv[][6]={0.1,0.3,0.2,0.0,0.7,0.5,0.6,0.0,0.3,0.4,0.1,0.2};
float outv[][4]={0.4,0.2,0.1,0.0};

cout<<"\n\nTHIS PROGRAM IS FOR A FUZZY ASSOCIATIVE MEMORY \
    NETWORK. THE NETWORK \n";
```

```
cout<<"IS SET UP FOR ILLUSTRATION WITH "<<ar<<" INPUT \
    NEURONS, AND "<<br;
cout<<" OUTPUT NEURONS.\n"<<nex<<" exemplars are used to \
    encode \n";

static network famn;
famn.getnwk(ar,br,nex,inptv,outv);
famn.iterate();
famn.findassc(inptv[1]);
famn.prstatus();

}
```

Output

The illustrative run of the previous program uses the fuzzy sets with fit vectors (0.1, 0.3, 0.2, 0.0, 0.7, 0.5) and (0.4, 0.2, 0.1, 0.0). As you can expect according to the theorem cited earlier, recall is not perfect in the reverse direction and the fuzzy associated memory consists of the pair (0.1, 0.3, 0.2, 0.0, 0.4, 0.4), and (0.4, 0.2, 0.1, 0.0). The computer output is in bold face and is in such detail as to be self-explanatory.

THIS PROGRAM IS FOR A FUZZY ASSOCIATIVE MEMORY NETWORK. THE NETWORK IS SET UP FOR ILLUSTRATION WITH SIX INPUT NEURONS, AND FOUR OUTPUT NEURONS.
I exemplars are used to encode

X vector you gave is:
0.I 0.3 0.2 0 0.7 0.5
Y vector you gave is:
0.4 0.2 0.I 0

weights--input layer to output layer:

0.I 0.I 0.I 0
0.3 0.2 0.I 0
0.2 0.2 0.I 0
0 0 0 0
0.4 0.2 0.I 0
0.4 0.2 0.I 0

weights--output layer to input layer:

0.I 0.3 0.2 0 0.4 0.4

0.1 0.2 0.2 0 0.2 0.2
0.1 0.1 0.1 0 0.1 0.1
0 0 0 0 0 0

Input vector is:
0.1 0.3 0.2 0 0.7 0.5

output layer neuron 0 activation is 0.4

output layer neuron 0 output is 0.4

output layer neuron 1 activation is 0.2

output layer neuron 1 output is 0.2

output layer neuron 2 activation is 0.1

output layer neuron 2 output is 0.1

output layer neuron 3 activation is 0

 output layer neuron 3 output is 0

X vector in possible associated pair is:
0.1 0.3 0.2 0 0.7 0.5
Y vector in possible associated pair is:
0.4 0.2 0.1 0

input layer neuron 0 activation is 0.1

input layer neuron 0 output is 0.1

input layer neuron 1 activation is 0.3

input layer neuron 1 output is 0.3

input layer neuron 2 activation is 0.2

input layer neuron 2 output is 0.2

input layer neuron 3 activation is 0

input layer neuron 3 output is 0

input layer neuron 4 activation is 0.4

input layer neuron 4 output is 0.4

input layer neuron 5 activation is 0.4

input layer neuron 5 output is 0.4

output layer neuron 0 activation is 0.4

output layer neuron 0 output is 0.4

output layer neuron 1 activation is 0.2

output layer neuron 1 output is 0.2

output layer neuron 2 activation is 0.1

output layer neuron 2 output is 0.1

output layer neuron 3 activation is 0

output layer neuron 3 output is 0

X vector in possible associated pair is:
0.1 0.3 0.2 0 0.4 0.4
Y vector in possible associated pair is:
0.4 0.2 0.1 0

PATTERNS ASSOCIATED:

X vector in the associated pair no. 1 is:
0.1 0.3 0.2 0 0.4 0.4
Y vector in the associated pair no. 1 is:
0.4 0.2 0.1 0

Input vector is:
0.6 0 0.3 0.4 0.1 0.2

output layer neuron 0 activation is 0.2

output layer neuron 0 output is 0.2

output layer neuron 1 activation is 0.2

output layer neuron 1 output is 0.2

output layer neuron 2 activation is 0.1

output layer neuron 2 output is 0.1

output layer neuron 3 activation is 0

output layer neuron 3 output is 0

X vector in possible associated pair is:
0.6 0 0.3 0.4 0.1 0.2
Y vector in possible associated pair is:
0.2 0.2 0.1 0

input layer neuron 0 activation is 0.1

input layer neuron 0 output is 0.1

input layer neuron 1 activation is 0.2

input layer neuron 1 output is 0.2

input layer neuron 2 activation is 0.2

input layer neuron 2 output is 0.2

input layer neuron 3 activation is 0

input layer neuron 3 output is 0

input layer neuron 4 activation is 0.2

input layer neuron 4 output is 0.2

input layer neuron 5 activation is 0.2

input layer neuron 5 output is 0.2

output layer neuron 0 activation is 0.2

output layer neuron 0 output is 0.2

output layer neuron 1 activation is 0.2

output layer neuron 1 output is 0.2

output layer neuron 2 activation is 0.1

output layer neuron 2 output is 0.1

output layer neuron 3 activation is 0

output layer neuron 3 output is 0

X vector in possible associated pair is:
0.1 0.2 0.2 0 0.2 0.2
Y vector in possible associated pair is:
0.2 0.2 0.1 0

output layer neuron 0 activation is 0.2

output layer neuron 0 output is 0.2

output layer neuron 1 activation is 0.2

output layer neuron 1 output is 0.2

output layer neuron 2 activation is 0.1

output layer neuron 2 output is 0.1

output layer neuron 3 activation is 0

output layer neuron 3 output is 0

output layer neuron 0 activation is 0.2

output layer neuron 0 output is 0.2

output layer neuron 1 activation is 0.2

output layer neuron 1 output is 0.2

output layer neuron 2 activation is 0.1

output layer neuron 2 output is 0.1

output layer neuron 3 activation is 0

output layer neuron 3 output is 0

X vector in possible associated pair is:
0.1 0.2 0.2 0 0.2 0.2
Y vector in possible associated pair is:
0.2 0.2 0.1 0

PATTERNS ASSOCIATED:

X vector in the associated pair no. 2 is:
0.1 0.2 0.2 0 0.2 0.2
Y vector in the associated pair no. 2 is:
0.2 0.2 0.1 0

THE FOLLOWING ASSOCIATED PAIRS WERE FOUND BY FUZZY AM

X vector in the associated pair no. 1 is:
0.1 0.3 0.2 0 0.4 0.4
Y vector in the associated pair no. 1 is:
0.4 0.2 0.1 0

X vector in the associated pair no. 2 is:
0.1 0.2 0.2 0 0.2 0.2
Y vector in the associated pair no. 2 is:
0.2 0.2 0.1 0

Summary

In this chapter bidirectional associative memories are presented for fuzzy subsets. The development of these is largely due to Kosko. They share the feature of resonance between the two layers in the network with Adaptive Resonance Theory. Even though there are connections in both directions between neurons in the two layers, essentially only one weight matrix is involved. You use the transpose of this weight matrix for the connections in the opposite direction. When one input at one end leads to some output at the other, which in turn leads to output same as the previous input, resonance is reached and an associated pair is found. In the case of bidirectional fuzzy associative memories, one pair of fuzzy sets determines one fuzzy associative memory system. Fit vectors are used in max–min composition. Perfect recall in both directions is not the case unless the heights of both fit vectors are equal. Fuzzy associative memories can improve the performance of an expert system by allowing fuzzy rules and learning more by experience.

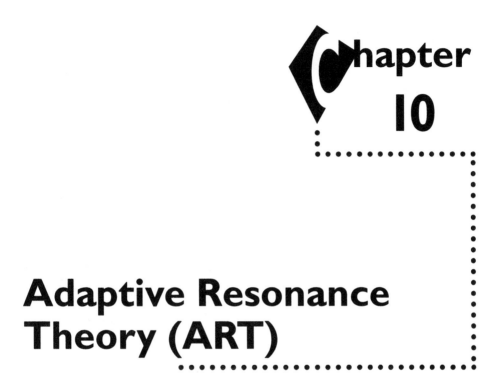

Chapter 10

Adaptive Resonance Theory (ART)

Introduction

The Adaptive Resonance Theory of Grossberg, developed further by Grossberg and Carpenter, is for the categorization of patterns using competitive learning paradigm. It introduces a gain control and a reset to make certain that learned categories are retained even while new categories are learned. It is successful in addressing the plasticity–stability dilemma.

Adaptive Resonance Theory makes much use of a competitive learning paradigm. A criterion is developed to facilitate the occurrence of winner-take-all phenomenon. A single node with the largest value for the set criterion is declared the winner within its layer, and it is said to classify a pattern class. If there is a tie for the winning neuron in a layer, then an arbitrary rule, such as the first of them in a serial order, can be taken as the winner.

The neural network developed for this theory establishes a system that is made up of two subsystems, one being the attentional subsystem, and this contains the unit for gain control. The other is an orienting subsystem, and this contains the unit for reset. During the operation of the network modeled for this theory, patterns emerge in the attentional subsystem and are called *traces of STM* or *short-term memory*. *Traces of LTM* (*long-term memory*) are in the connection weights between the input layer and output layer.

The network uses processing with feedback between its two layers, until resonance occurs. Resonance occurs when the output in the first layer after feedback from the second layer matches the original pattern used as input for the first layer in that processing cycle. A match of this type does not have to be perfect. What is required is that the degree of match, measured suitably, exceeds a predetermined level, termed the *vigilance parameter*. Just as a photograph matches the likeness of the subject to a greater degree when the granularity is higher, the pattern match gets finer when the vigilance parameter is closer to 1.

The Network for ART I

The neural network for the adaptive resonance theory or ART1 model consists of the following:

- ◆ A layer of neurons, called the F_1 layer (input layer or comparison layer)
- ◆ A node for each layer as a gain control unit
- ◆ A layer of neurons, called the F_2 layer (*output* layer or *recognition* layer)
- ◆ A node as a reset unit
- ◆ Bottom-up connections from F_1 layer to F_2 layer
- ◆ Top-down connections from F_2 layer to F_1 layer
- ◆ Inhibitory connection (negative weight) form F_2 layer to gain control
- ◆ Excitatory connection (positive weight) from gain control to a layer

- ◆ Inhibitory connection from F_1 layer to reset node
- ◆ Excitatory connection from reset node to F_2 layer

A Simplified Diagram of Network Layout

Figure 10.1 shows a simplified diagram of the neural network for an ART1 model.

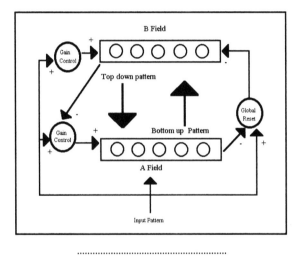

FIGURE 10.1   *ART1 Layout*

Processing in ART1

Special Features of the ART1 Model

One special feature of an ART1 model is that a so-called two-thirds rule is necessary to determine the activity of neurons in the F_1 layer. There would be three input sources to the neuron. They are the external input, the output of gain control, and the outputs of F_2 layer neurons as inputs to F_1 layer. The neuron will not fire unless at least two of the three inputs are active. The gain control unit and the two-thirds rule together assure proper response from the input layer neurons. A second feature is that a vigilance parameter is used to determine the activity of the reset unit.

Notation for ART1 Calculations

Let us list the various symbols we will use to describe the operation of a neural network for an ART1 model.

$\mathbf{w_{ij}}$	weight on the connection from the ith neuron in the F_1 layer to the jth neuron in the F_2 layer
$\mathbf{v_{ji}}$	weight on the connection from the jth neuron in the F_2 layer to the ith neuron on the F_1 layer
$\mathbf{a_i}$	activation of ith neuron in the F_1 layer
$\mathbf{b_j}$	activation of jth neuron in the F_2 layer
$\mathbf{x_i}$	output of ith neuron in the F_1 layer
$\mathbf{y_j}$	output of jth neuron in the F_2 layer
$\mathbf{z_i}$	input to ith neuron in F_1 layer from F_2 layer
ρ	vigilance parameter, positive and no greater than 1
$\mathbf{m}$	number of neurons in the F_1 layer
$\mathbf{n}$	number of neurons in the F_2 layer
$\mathbf{I}$	input vector
$\mathbf{S_i}$	sum of the components of the input vector
$\mathbf{S_x}$	sum of the outputs of neurons in the F_1 layer
$\mathbf{A, C, D}$	parameters with positive values or zero
$\mathbf{L}$	parameter with value greater than 1
$\mathbf{B}$	parameter with value less than $D + 1$ but at least as large as either D or 1
$\mathbf{r}$	index of winner of competition in the F_2 layer

Algorithm for ART1 Calculations

The ART1 equations are not easy to follow. We follow the description of the algorithm found in James A. Freeman and David M. Skapura. The following equations, taken in the order given, describe the steps in the algorithm. Note that **binary** input patterns are used in ART1.

Initialization of Parameters

```
wij should be positive and less than  L / ( m - 1 + L)
vji should be greater than ( B - 1 ) / D
ai = -B / ( 1 + C )
```

Equations for ART I Computations

When you read below the equations for ART1 computations, keep in mind the following considerations. If a subscript i appears on the left-hand side of the equation, it means that there are m such equations, as the subscript i varies from 1 to m. Similarly, if instead a subscript j occurs, then there are n such equations as j ranges from 1 to n. The equations are used in the order they are given. They give a step by step description of the following algorithm. All the variables, you recall, are defined in the earlier section on notation. For example, I is the input vector.

F₁ layer calculations:
```
ai = Ii / ( 1+ A ( Ii + B ) + C )
xi =  1 if  ai > 0
      0 if  ai ≤ 0
```

F₂ layer calculations:
```
bj = Σ wij xi, the summation being on i from 1 to m
yj = 1 if jth neuron has the largest activation value in the
        F2 layer
     0 if jth neuron is not the winner in F2 layer
```

Top-down inputs:
```
zi = Σvjiyj, the summation being on j from 1 to n (You will
             notice that all but one terms are zero)
```

F₁ layer calculations:
```
ai = ( Ii + D zi - B ) / ( 1 + A ( Ii + D zi ) + C )
xi =  1 if  ai > 0
      0 if  ai ≤ 0
```

Checking with vigilance parameter:
```
If ( Sx / SI ) <ρ, set yj = 0 for all j, including the win-
```

```
ner r in F2 layer, and consider the jth neuron inactive (this
step is reset)
If  ( Sx / SI ) ≥ ρ, then
```

Modifying top-down and bottom-up connection weight for winner r:

$$v_{ir} = (L / (Sx + L -1) \text{ if } x_i = 1$$
$$\qquad 0 \quad \text{if } x_i = 0$$
$$w_{ri} = 1 \text{ if } x_i = 1$$
$$\qquad 0 \text{ if } x_i = 0$$

Having finished with the current input pattern, we repeat these steps with a new input pattern. We lose the index r given to one neuron as a winner and treat all neurons in the F_2 layer with their original indices (subscripts).

The above presentation of the algorithm is hoped to make all the steps as clear as possible. The process is rather involved. To recapitulate, first an input vector is presented to the F_1 layer neurons, their activations are determined, and then the **threshold** function is used. The outputs of the F_1 layer neurons constitute the inputs to the F_2 layer neurons, from which a winner is designated on the basis of the largest activation. The winner only is allowed to be active, meaning that the output is 1 for the winner and 0 for all the rest. The equations implicitly incorporate the use of the 2/3 rule that we mentioned earlier, and they also incorporate the way the gain control is used. The gain control is designed to have a value 1 in the phase of determining the activations of the neurons in the F_2 layer and 0 if either there is no input vector or output from the F_2 layer is backpropagated to the F_1 layer.

Other Models

Extensions of an ART1 model, which is for binary patterns, are ART2 and ART3. Of these, ART2 model categorizes and stores analog-valued patterns, as well as binary patterns, while ART3 addresses computational problems of hierarchies.

C++ Implementation

Again, the algorithm for ART1 processing as given in Freeman and Skapura is followed for our C++ implementation. We may point out that as you look at different C++ programs, you will notice some reuse of the code, even if the models for which the programs are written differ. Such reuse, as mentioned early in the book, makes C++ and other object-oriented programming languages attractive.

A Header File for the C++ Program for the ART1 Model Network

The header file for the C++ program for the ART1 model network is *art1net.hpp*. It contains the declarations for two classes, an **artneuron** class for neurons in the ART1 model, and a **network** class, which is declared as a **friend** class in the **artneuron** class. Functions declared in the **network** class include one to do the iterations for the network operation, finding the winner in a given iteration, and one to inquire if reset is needed.

```
//art1net.h    V. Rao,  H. Rao
//Header file for ART1 model network program

#include <iostream.h>
#define MXSIZ 10


class artneuron
{

protected:
    int nnbr;
    int inn,outn;
    int output;
    double activation;
    double outwt[MXSIZ];
    char *name;
    friend class network;

public:
    artneuron() { };
    void getnrn(int,int,int,char *);
};

class network
{
public:
    int   anmbr,bnmbr,flag,ninpt,sj,so,winr;
    float ai,be,ci,di,el,rho;
    artneuron (anrn)[MXSIZ],(bnrn)[MXSIZ];
```

```
int outs1[MXSIZ],outs2[MXSIZ];
int lrndptrn[MXSIZ][MXSIZ];
double acts1[MXSIZ],acts2[MXSIZ];
double mtrx1[MXSIZ][MXSIZ],mtrx2[MXSIZ][MXSIZ];

network() { };
void getnwk(int,int,float,float,float,float,float);
void prwts1();
void prwts2();
int winner(int k,double *v,int);
void practs1();
void practs2();
void prouts1();
void prouts2();
void iterate(int *,float,int);
void asgninpt(int *);
void comput1(int);
void comput2(int *);
void prlrndp();
void inqreset(int);
void adjwts1();
void adjwts2();
};
```

A Source File for C++ Program for an ART I Model Network

The implementations of the functions declared in the header file are contained in the source file for the C++ program for an ART1 model network. It also has the **main** function, which contains specifications of the number of neurons in the two layers of the network, the values of the vigilance and other parameters, and the input vectors. Note that if there are n neurons in a layer, they are numbered serially from 0 to n–1, and not from 1 to n in the C++ program. The source file is called *art1net.cpp*. It is set up with six neurons in the F_1 layer and seven neurons in the F_2 layer. The **main** function also contains the parameters needed in the algorithm.

To initialize the bottom-up weights, we set each weight to be –0.1 + L/(m – 1 + L) so that it is greater than 0 and less than L/(m – 1 + L), as suggested before. Similarly, the top-down weights are initialized by setting each of them to 0.2 + (B – 1)/D so it would be greater than (B – 1)/D. Initial activations of the F_1 layer neurons are each set to –B/(1 + C), as suggested earlier.

A **restrmax** function is defined to compute the maximum in an array when one of the array elements is not desired to be a candidate for the maximum. This facilitates the removal of the current winner from competition when reset is needed. Reset is needed when the degree of match is of a smaller magnitude than the vigilance parameter.

The function *iterate* is a **member** function of the network class and does the processing for the network. The **inqreset** function of the **network** class compares the vigilance parameter with the degree of match.

```
//art1net.cpp  V. Rao, H. Rao
//Source file for ART1 network program

#include "art1net.h"

int restrmax(int j,double *b,int k)
    {
    int i,tmp;

    for(i=0;i<j;i++){
        if(i !=k)
        {tmp = i;
        i = j;}
        }

    for(i=0;i<j;i++){

    if( (i != tmp)&&(i != k))

        {if(b[i]>b[tmp]) tmp = i;}}

    return tmp;
    }
void artneuron::getnrn(int m1,int m2,int m3, char *y)
    {
int i;
name = y;
nnbr = m1;
outn = m2;
inn  = m3;

for(i=0;i<outn;++i){

    outwt[i] = 0 ;
    }

output = 0;
activation = 0.0;
```

```
}

    void network::getnwk(int k,int l,float aa,float bb,float
    cc,float dd,float ll)
{
anmbr = k;
bnmbr = l;
ninpt = 0;
ai = aa;
be = bb;
ci = cc;
di = dd;
el = ll;
int i,j;
flag = 0;

char *y1="ANEURON", *y2="BNEURON" ;

for(i=0;i<anmbr;++i){

    anrn[i].artneuron::getnrn(i,bnmbr,0,y1);}

for(i=0;i<bnmbr;++i){

    bnrn[i].artneuron::getnrn(i,0,anmbr,y2);}

float tmp1,tmp2,tmp3;
tmp1 = 0.2 +(be - 1.0)/di;
tmp2 = -0.1 + el/(anmbr - 1.0 +el);
tmp3 = - be/(1.0 + ci);

for(i=0;i<anmbr;++i){

    anrn[i].activation = tmp3;
    acts1[i] = tmp3;

    for(j=0;j<bnmbr;++j){

        mtrx1[i][j]  = tmp1;
        mtrx2[j][i] = tmp2;
        anrn[i].outwt[j] = mtrx1[i][j];
        bnrn[j].outwt[i] = mtrx2[j][i];
        }
    }

prwts1();
prwts2();
practs1();
cout<<"\n";
}
```

```
int network::winner(int k,double *v,int kk){
int t1;

t1 = restrmax(k,v,kk);
return t1;
}

void network::prwts1()
{
int i3,i4;
cout<<"\nweights for F1 layer neurons: \n";

for(i3=0;i3<anmbr;++i3){

    for(i4=0;i4<bnmbr;++i4){

       cout<<anrn[i3].outwt[i4]<<"  ";}

    cout<<"\n"; }

cout<<"\n";
}

void network::prwts2()
{
int i3,i4;
cout<<"\nweights for F2 layer neurons: \n";

for(i3=0;i3<bnmbr;++i3){

    for(i4=0;i4<anmbr;++i4){

       cout<<bnrn[i3].outwt[i4]<<"  ";};

    cout<<"\n";   }

cout<<"\n";
}

void network::practs1()
{
int j;
cout<<"\nactivations of F1 layer neurons: \n";

for(j=0;j<anmbr;++j){

    cout<<acts1[j]<<"   ";}

cout<<"\n";
}
```

```
void network::practs2()
{
int j;
cout<<"\nactivations of F2 layer neurons: \n";

for(j=0;j<bnmbr;++j){

    cout<<acts2[j]<<"    ";}

cout<<"\n";
}

void network::prouts1()
{
int j;
cout<<"\noutputs of F1 layer neurons: \n";

for(j=0;j<anmbr;++j){

    cout<<outs1[j]<<"    ";}

cout<<"\n";
}

void network::prouts2()
{
int j;
cout<<"\noutputs of F2 layer neurons: \n";

for(j=0;j<bnmbr;++j){

    cout<<outs2[j]<<"    ";}

cout<<"\n";
}

void network::asgninpt(int *b)
{
int j;
sj = so = 0;
cout<<"\nInput vector is:\n" ;

for(j=0;j<anmbr;++j){

    cout<<b[j]<<" ";}

cout<<"\n";

for(j=0;j<anmbr;++j){

    sj += b[j];
```

```
        anrn[j].activation = b[j]/(1.0 +ci +ai*(b[j]+be));
        acts1[j] = anrn[j].activation;

        if(anrn[j].activation > 0) anrn[j].output = 1;

        else
            anrn[j].output = 0;

        outs1[j] = anrn[j].output;
        so += anrn[j].output;
        }

practs1();
prouts1();
}

void network::inqreset(int t1)
{
int jj;
flag = 0;
jj = so/sj;
cout<<"\ndegree of match: "<<jj<<" vigilance:  "<<rho<<"\n";

if( jj > rho ) flag = 1;

    else
    {cout<<"winner is "<<t1;
    cout<<" reset required \n";}

}

void network::comput1(int k)
{
int j;

for(j=0;j<bnmbr;++j){

    int ii1;
    double c1 = 0.0;
    cout<<"\n";

    for(ii1=0;ii1<anmbr;++ii1){

        c1 += outs1[ii1] * mtrx2[j][ii1];
        }

    bnrn[j].activation = c1;
    acts2[j] = c1;};

winr = winner(bnmbr,acts2,k);
```

```
cout<<"winner is "<<winr;
for(j=0;j<bnmbr;++j){

    if(j == winr) bnrn[j].output = 1;

    else bnrn[j].output =  0;
    outs2[j] = bnrn[j].output;
    }

practs2();
prouts2();
}

void network::comput2(int *b)
{
double db[MXSIZ];
double tmp;
so = 0;
int i,j;

for(j=0;j<anmbr;++j){

    db[j] =0.0;

    for(i=0;i<bnmbr;++i){

        db[j] += mtrx1[j][i]*outs2[i];};

    tmp = b[j] + di*db[j];
    acts1[j] = (tmp - be)/(ci +1.0 +ai*tmp);
    anrn[j].activation = acts1[j];

    if(anrn[j].activation > 0) anrn[j].output = 1;

    else anrn[j].output = 0;

    outs1[j] = anrn[j].output;
    so += anrn[j].output;
    }
cout<<"\n";
practs1();
prouts1();
}

void network::adjwts1()
{
int i;

for(i=0;i<anmbr;++i){
```

```
        if(outs1[i] >0) {mtrx1[i][winr]  = 1.0;}

        else

            {mtrx1[i][winr] = 0.0;}

        anrn[i].outwt[winr] = mtrx1[i][winr];}
prwts1();
}
void network::adjwts2()
{
int i;
cout<<"\nwinner is "<<winr<<"\n";

for(i=0;i<anmbr;++i){

        if(outs1[i] > 0) {mtrx2[winr][i] = el/(so + el -1);}

        else

            {mtrx2[winr][i] = 0.0;}

        bnrn[winr].outwt[i]  = mtrx2[winr][i];}

prwts2();
}

void network::iterate(int *b,float rr,int kk)
{
int j;
rho = rr;
flag = 0;

asgninpt(b);
comput1(kk);
comput2(b);
inqreset(winr);

if(flag == 1){

    ninpt ++;
    adjwts1();
    adjwts2();
    int j3;

    for(j3=0;j3<anmbr;++j3){

        lrndptrn[ninpt][j3] = b[j3];}
```

```
        prlrndp();
        }

    else

        {

        for(j=0;j<bnmbr;++j){

            outs2[j] = 0;
            bnrn[j].output = 0;}

        iterate(b,rr,winr);
        }
}

void network::prlrndp()
{
int j;
cout<<"\nlearned vector # "<<ninpt<<"   :\n";

for(j=0;j<anmbr;++j){

    cout<<lrndptrn[ninpt][j]<<"  ";}

cout<<"\n";
}

void main()
{
int ar = 6, br = 7, rs = 8;
float aa = 2.0,bb = 2.5,cc = 6.0,dd = 0.85,ll = 4.0,rr =
    0.95;
int inptv[][6]={0,1,0,0,0,0,1,0,1,0,1,0,0,0,0,0,0,1,0,1,0,1,0,\
    1,0};

cout<<"\n\nTHIS PROGRAM IS FOR AN -ADAPTIVE RESONANCE THEORY\
    1 - NETWORK.\n";
cout<<"THE NETWORK IS SET UP FOR ILLUSTRATION WITH "<<ar<<" \
    INPUT NEURONS,\n";
cout<<" AND "<<br<<" OUTPUT NEURONS.\n";

static network bpn;
bpn.getnwk(ar,br,aa,bb,cc,dd,ll) ;
bpn.iterate(inptv[0],rr,rs);
bpn.iterate(inptv[1],rr,rs);
bpn.iterate(inptv[2],rr,rs);
bpn.iterate(inptv[3],rr,rs);
}
```

Output

The computer output is presented in boldface. Four input vectors are used in the trial run of the program and these are specified in the main function. The output is self-explanatory. We have included only in this text some comments regarding the output. These comments are enclosed within strings of asterisks. They are not actually part of the program output, and hence, they are not in boldface. Here is a summarization of the categorization of the inputs done by the network. Keep in mind that the numbering of the neurons in any layer, which has n neurons, is from 0 to n − 1, and not from 1 to n.

input	winner in F_2 layer
0 1 0 0 0 0	0, no reset
1 0 1 0 1 0	1, no reset
0 0 0 0 1 0	1, after reset 2
1 0 1 0 1 0	1, after reset 3

The input pattern 0 0 0 0 1 0 is considered a subset of the pattern 1 0 1 0 1 0 in the sense that in whatever position the first pattern has a 1, the second pattern also has a 1. Of course, the second pattern has 1's in other positions as well. At the same time, the pattern 1 0 1 0 1 0 is considered a superset of the pattern 0 0 0 0 1 0. The reason that the pattern 1 0 1 0 1 0 is repeated as input after the pattern 0 0 0 0 1 0 is processed, is to see what happens with this superset. In both cases, the degree of match falls short of the vigilance parameter, and a reset is needed.

THIS PROGRAM IS FOR AN ADAPTIVE RESONANCE THEORY I-NETWORK. THE NETWORK IS SET UP FOR ILLUSTRATION WITH SIX INPUT NEURONS AND SEVEN OUTPUT NEURONS.
****
Initialization of connection weights and F_1 layer activations. F_1 layer connection weights are all chosen to be equal to a random value subject to the conditions given in the algorithm. Similarly, F_2 layer connection weights are all chosen to be equal to a random value subject to the conditions given in the algorithm.
****
weights for F1 layer neurons:

```
1.964706  1.964706  1.964706  1.964706  1.964706  1.964706 1. 964706
1.964706  1.964706  1.964706  1.964706  1.964706  1.964706  1.964706
1.964706  1.964706  1.964706  1.964706  1.964706  1.964706  1.964706
1.964706  1.964706  1.964706  1.964706  1.964706  1.964706  1.964706
1.964706  1.964706  1.964706  1.964706  1.964706  1.964706  1.964706
1.964706  1.964706  1.964706  1.964706  1.964706  1.964706  1.964706
```

weights for F2 layer neurons:
```
0.344444  0.344444  0.344444  0.344444  0.344444  0.344444
0.344444  0.344444  0.344444  0.344444  0.344444  0.344444
0.344444  0.344444  0.344444  0.344444  0.344444  0.344444
0.344444  0.344444  0.344444  0.344444  0.344444  0.344444
0.344444  0.344444  0.344444  0.344444  0.344444  0.344444
0.344444  0.344444  0.344444  0.344444  0.344444  0.344444
0.344444  0.344444  0.344444  0.344444  0.344444  0.344444
```

activations of F1 layer neurons:
-0.357143 -0.357143 -0.357143 -0.357143 -0.357143 -0.357143
```
*****************************************************************
A new input vector and a new iteration
*****************************************************************
```
Input vector is:
0 1 0 0 0 0

activations of F1 layer neurons:
0 0.071429 0 0 0 0

outputs of F1 layer neurons:
0 1 0 0 0 0

winner is 0
activations of F2 layer neurons:
0.344444 0.344444 0.344444 0.344444 0.344444 0.344444 0.344444

outputs of F2 layer neurons:
1 0 0 0 0 0

activations of F1 layer neurons:
-0.080271 0.013776 -0.080271 -0.080271 -0.080271 -0.080271

outputs of F1 layer neurons:
0 1 0 0 0 0
```
*****************************************************************
```
Top-down and bottom-up outputs at F_1 layer match, showing resonance.
```
*****************************************************************
```
degree of match: 1 vigilance: 0.95

weights for F1 layer neurons:
0 1.964706 1.964706 1.964706 1.964706 1.964706 1.964706

```
1 1.964706  1.964706  1.964706  1.964706  1.964706  1.964706
0 1.964706  1.964706  1.964706  1.964706  1.964706  1.964706
0 1.964706  1.964706  1.964706  1.964706  1.964706  1.964706
0 1.964706  1.964706  1.964706  1.964706  1.964706  1.964706
0 1.964706  1.964706  1.964706  1.964706  1.964706  1.964706
```

winner is 0

weights for F2 layer neurons:
```
0 1 0 0 0 0
0.344444 0.344444 0.344444 0.344444 0.344444 0.344444
0.344444 0.344444 0.344444 0.344444 0.344444 0.344444
0.344444 0.344444 0.344444 0.344444 0.344444 0.344444
0.344444 0.344444 0.344444 0.344444 0.344444 0.344444
0.344444 0.344444 0.344444 0.344444 0.344444 0.344444
0.344444 0.344444 0.344444 0.344444 0.344444 0.344444
```

learned vector # 1 :
```
0 1 0 0 0 0
**********************************************************
A new input vector and a new iteration
**********************************************************
```
Input vector is:
```
1 0 1 0 1 0
```

activations of F1 layer neurons:
```
0.071429 0 0.071429 0 0.071429 0
```

outputs of F1 layer neurons:
```
1 0 1 0 1 0
```

winner is 1
activations of F2 layer neurons:
```
0 1.033333  1.033333  1.033333  1.033333  1.033333  1.033333
```

outputs of F2 layer neurons:
```
0 1 0 0 0 0 0
```

activations of F1 layer neurons:
```
0.013776  -0.080271  0.013776  -0.080271  0.013776  -0.080271
```

outputs of F1 layer neurons:
```
1 0 1 0 1 0
**********************************************************
Top-down and bottom-up outputs at F₁ layer match,
showing resonance.
**********************************************************
```
degree of match: 1 vigilance: 0.95

weights for F1 layer neurons:
```
0 1 1.964706  1.964706  1.964706  1.964706  1.964706
```

```
1  0  1.964706  1.964706  1.964706  1.964706  1.964706
0  1  1.964706  1.964706  1.964706  1.964706  1.964706
0  0  1.964706  1.964706  1.964706  1.964706  1.964706
0  1  1.964706  1.964706  1.964706  1.964706  1.964706
0  0  1.964706  1.964706  1.964706  1.964706  1.964706
```

winner is 1

weights for F2 layer neurons:
```
0  1  0  0  0  0
0.666667  0  0.666667  0  0.666667  0
0.344444  0.344444  0.344444  0.344444  0.344444  0.344444
0.344444  0.344444  0.344444  0.344444  0.344444  0.344444
0.344444  0.344444  0.344444  0.344444  0.344444  0.344444
0.344444  0.344444  0.344444  0.344444  0.344444  0.344444
0.344444  0.344444  0.344444  0.344444  0.344444  0.344444
```

learned vector # 2 :
```
1  0  1  0  1  0
```
```
*****************************************************************
A new input vector and a new iteration
*****************************************************************
```
Input vector is:
```
0  0  0  0  1  0
```

activations of F1 layer neurons:
```
0  0  0  0  0.071429  0
```

outputs of F1 layer neurons:
```
0  0  0  0  1  0
```

winner is 1
activations of F2 layer neurons:
```
0  0.666667  0.344444  0.344444  0.344444  0.344444  0.344444
```

outputs of F2 layer neurons:
```
0  1  0  0  0  0
```

activations of F1 layer neurons:
```
-0.189655  -0.357143  -0.189655  -0.357143  -0.060748  -0.357143
```

outputs of F1 layer neurons:
```
0  0  0  0  0  0
```

degree of match: 0 vigilance: 0.95
winner is 1 reset required
```
*****************************************************************
Input vector repeated after reset, and a new iteration
*****************************************************************
```
Input vector is:

0 0 0 0 I 0

activations of FI layer neurons:
0   0   0   0   0.071429   0

outputs of FI layer neurons:
0   0   0   0   I   0

winner is 2
activations of F2 layer neurons:
0   0.666667   0.344444   0.344444   0.344444   0.344444   0.344444
outputs of F2 layer neurons:
0   0   I   0   0   0   0

activations of FI layer neurons:
-0.080271   -0.080271   -0.080271   -0.080271   0.013776   -0.080271

outputs of FI layer neurons:
0   0   0   0   I   0
```
****************************************************************
```
Top-down and bottom-up outputs at F_1 layer match, showing
resonance.
```
****************************************************************
```
degree of match: I vigilance: 0.95

weights for FI layer neurons:
0   I   0   1.964706   1.964706   1.964706   1.964706
I   0   0   1.964706   1.964706   1.964706   1.964706
0   I   0   1.964706   1.964706   1.964706   1.964706
0   0   0   1.964706   1.964706   1.964706   1.964706
0   I   I   1.964706   1.964706   1.964706   1.964706
0   0   0   1.964706   1.964706   1.964706   1.964706

winner is 2

weights for F2 layer neurons:
0   I   0   0   0   0
0.666667   0   0.666667   0   0.666667   0
0   0   0   0   I   0
0.344444   0.344444   0.344444   0.344444   0.344444   0.344444
0.344444   0.344444   0.344444   0.344444   0.344444   0.344444
0.344444   0.344444   0.344444   0.344444   0.344444   0.344444
0.344444   0.344444   0.344444   0.344444   0.344444   0.344444

learned vector # 3 :
0   0   0   0   I   0
```
****************************************************************
```
An old (actually the second above) input vector is retried
after trying a subset vector, and a new iteration
```
****************************************************************
```
Input vector is:

```
1 0 1 0 1 0
```

activations of F1 layer neurons:
0.071429 0 0.071429 0 0.071429 0

outputs of F1 layer neurons:
1 0 1 0 1 0

winner is 1
activations of F2 layer neurons:
0 2 1 1.033333 1.033333 1.033333 1.03333

outputs of F2 layer neurons:
0 1 0 0 0 0 0

activations of F1 layer neurons:
-0.060748 -0.357143 -0.060748 -0.357143 -0.060748 -0.357143

outputs of F1 layer neurons:
0 0 0 0 0 0

degree of match: 0 vigilance: 0.95
winner is 1 reset required
****
Input vector repeated after reset, and a new iteration
****
Input vector is:
1 0 1 0 1 0

activations of F1 layer neurons:
0.071429 0 0.071429 0 0.071429 0

outputs of F1 layer neurons:
1 0 1 0 1 0

winner is 3
activations of F2 layer neurons:
0 2 1 1.033333 1.033333 1.033333 1.033333

outputs of F2 layer neurons:
0 0 0 1 0 0 0

activations of F1 layer neurons:
0.013776 -0.080271 0.013776 -0.080271 0.013776 -0.080271

outputs of F1 layer neurons:
1 0 1 0 1 0
****
Top-down and Bottom-up outputs at F1layer match, showing res-
onance.
****

degree of match: 1 vigilance: 0.95

weights for F1 layer neurons:
```
0  1  0  1  1.964706  1.964706  1.964706
1  0  0  0  1.964706  1.964706  1.964706
0  1  0  1  1.964706  1.964706  1.964706
0  0  0  0  1.964706  1.964706  1.964706
0  1  1  1  1.964706  1.964706  1.964706
0  0  0  0  1.964706  1.964706  1.964706
```

winner is 3

weights for F2 layer neurons:
```
0  1  0  0  0  0
0.666667  0  0.666667  0  0.666667  0
0  0  0  0  1  0
0.666667  0  0.666667  0  0.666667  0
0.344444  0.344444  0.344444  0.344444  0.344444  0.344444
0.344444  0.344444  0.344444  0.344444  0.344444  0.344444
0.344444  0.344444  0.344444  0.344444  0.344444  0.344444
```

learned vector # 4 :
```
1  0  1  0  1  0
```

Summary

This chapter presented the basics of the Adaptive Resonance Theory of Grossberg and Carpenter and the C++ implementation of the neural network modeled for this theory. It is an elegant theory and it addresses the stability–plasticity dilemma. The network relies on resonance. It is a self-organizing network and does categorization by associating individual neurons of the F_2 layer with individual patterns. By employing a so-called 2/3 rule, it ensures stability in learning patterns.

Chapter
11

The Kohonen Self-Organizing Map

Introduction

This chapter will discuss one type of unsupervised competitive learning, using what is called a *Kohonen feature map*, or *Self-Organizing Map* (SOM). As you recall, in unsupervised learning there are no expected outputs presented to a neural network, as in the case of a supervised training algorithm such as backpropagation. Instead, a network, by its self-organizing properties, is able to infer relationships and learn more as more inputs are presented to it. One advantage to this scheme is that you can expect the system to change with changing conditions and inputs. The system constantly learns. The Kohonen SOM is a neural network system developed by Teuvo Kohonen of Helsinki University of Technology and is often used to classify inputs into different

223

categories. Applications for feature maps can be traced to many areas, including speech recognition and robot motor control.

Competitive Learning

A Kohonen feature map may be used by itself or as a layer of another neural network. A Kohonen layer is composed of neurons that compete with each other. It is another case of a winner-take-all strategy. Inputs are fed into each of the neurons in the Kohonen layer (from the input layer). Each neuron determines its output according to a weighted sum formula:

```
Output  = Σ wij xi
```

The weights and the inputs are usually normalized, which means that the magnitude of the weight and input vectors are set equal to one. The neuron with the largest output is the winner. This neuron has a final output of 1. All other neurons in the layer have an output of zero. Input patterns end up firing different winner neurons. This classifies similar or identical input patterns to an output neuron. In Chapter 12, you will see the use of a Kohonen network in pattern classification.

Normalization of a Vector

Consider a vector, $A = ax + by + cz$. The normalized vector A' is obtained by dividing each component of A by the square root of the sum of squares of all the components. In other words each component is multiplied by $1/ \sqrt{(a^2 + b^2 + c^2)}$. Both the weight vector and the input vector are normalized during the operation of the Kohonen feature map. The reason for this is that the training law uses subtraction of the weight vector from the input vector. Using normalization of the values in the subtraction reduces both vectors to a unit-less status, and hence, makes the subtraction of like quantities possible. You will learn more about the training law shortly.

Lateral Inhibition

Lateral inhibition is a process that takes place in some biological neural networks. Lateral connections of neurons in a given layer are formed, and squash distant neighbors. The strength of connections is inversely related to distance. The positive, supportive connections are termed as *excitatory* while the negative, squashing connections are termed *inhibitory*.

A biological example of lateral inhibition occurs in the human vision system.

The Mexican Hat Function

Figure 11.1 shows a function, called the **mexican hat** function, which shows the relationship between the connection strength and the distance from the winning neuron. The effect of this function is to set up a competitive environment for learning. Only winning neurons and their neighbors participate in learning for a given input pattern.

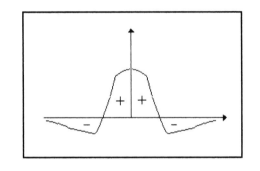

FIGURE 11.1   *The **mexican hat** function showing lateral inhibition.*

Training Law for the Kohonen Map

The training law for the Kohonen feature map is straightforward. The change in weight vector for a given output neuron is a gain constant, alpha, multiplied by the difference between the input vector and the old weight vector:

```
Wnew = Wold +  alpha * (Input -Wold)
```

Both the old weight vector and the input vector are normalized to unit length. Alpha is a gain constant between 0 and 1.

Significance of the Training Law

Let us consider the case of a two-dimensional input vector. If you look at a unit circle, as shown in Figure 11.2, the effect of the training law is to try to align the weight vector and the input vector. Each pattern attempts to nudge the weight vector closer by a fraction determined by alpha. For three dimensions the surface becomes a unit sphere instead of a circle. For higher dimensions you term the surface a *hypersphere*. It is not necessarily ideal to have perfect alignment of the input and weight vectors. You use neural networks for their ability to recognize patterns, but also to generalize input data sets. By aligning all input vectors to the corresponding winner weight vectors, you are essentially *memorizing* the input data set classes. It may be more desirable to come close, so that noisy or incomplete inputs may still trigger the correct classification.

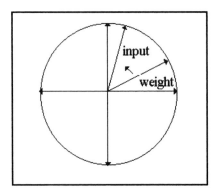

FIGURE 11.2 *The training law for the Kohonen map as shown on a unit circle.*

The Neighborhood Size and Alpha

In the Kohonen map, a parameter called the *neighborhood size* is used to model the effect of the **mexican hat** function. Those neurons that are within the distance specified by the neighborhood size participate in training and weight vector changes; those that are outside this distance do not participate in learning. The neighborhood size typically is started as an initial value and is decreased as the input pattern cycles continue. This process tends to support the winner-take-all strategy by eventually singling out a winner neuron for a given pattern.

Figure 11.3 shows a linear arrangement of neurons with a neighborhood size of 2. The hashed central neuron is the winner. The darkened adjacent neurons are those that will participate in training.

Besides the neighborhood size, alpha typically is also reduced during simulation. You will see these features when we develop a Kohonen map program.

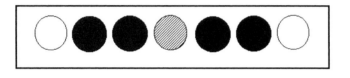

FIGURE 11.3   *Winner neuron with a neighborhood size of 2 for a Kohonen map.*

C++ Code for Implementing a Kohonen Map

The C++ code for the Kohonen map draws on much of the code developed for the backpropagation simulator. The Kohonen map is a much simpler program and does not rely on large data sets for input. The Kohonen map program uses only two files, an input file and an output file. In order to use the program, you must create an input data set and save this in a file called *input.dat*. The output file is called *kohonen.dat* and is saved in your current working directory. You will get more details shortly on the formats of these files.

The Kohonen Network

The Kohonen network has two layers, an *input* layer and a *Kohonen output* layer. This is shown in Figure 11.4. The input layer is a size determined by the user and must match the size of each row (pattern) in the input data file.

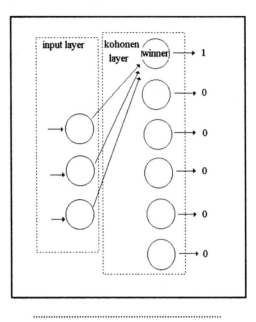

FIGURE 11.4 *A Kohonen network.*

Modeling Lateral Inhibition

For the sake of computational efficiency, you will see that lateral inhibition is modeled by looking at the maximum output for the output neurons and making that output belong to a winner neuron. Other outputs are inhibited by setting their outputs to zero. The true way of modeling lateral inhibition would be too expensive since the number of lateral connections is quite large. Training, or weight update, is performed on all outputs that are within a neighborhood size distance from the winner neuron.

Classes to be Used

We use many of the classes from the backpropagation simulator. We require only two layers , the input layer and the Kohonen layer. We make a new layer class called the **Kohonen layer** class, and a new network class called the **Kohonen_network**.

Revisiting the Layer Class

The **layer** class needs to be slightly modified, as shown in Listing 11.1.

```
Listing 11.1        Modification of layer.h
// layer.h    V.Rao, H. Rao
// header file for the layer class hierarchy and
// the network class

#define MAX_LAYERS  5
#define MAX_VECTORS 100


class network;
class Kohonen_network;

class layer
{

protected:

    int num_inputs;
    int num_outputs;
    float *outputs;  // pointer to array of outputs
    float *inputs;   // pointer to array of inputs, which
                // are outputs of some other layer

    friend network;
    friend Kohonen_network; // update for Kohonen model

public:

    virtual void calc_out()=0;
};
...
```

Here the changes are indicated in boldface. You notice that the **Kohonen_network** is made a friend to the **layer** class, so that the **Kohonen_network** can have access to the data of a layer.

A New Layer Class for a Kohonen Layer

The next step to take is to create a **Kohonen_layer** class and a **Kohonen_network** class. This is shown in Listing 11.2.

```
Listing 11.2   The Kohonen_layer class and Kohonen_network class in layerk.h
// layerk.h     V.Rao, H. Rao
```

```
// header file for the Kohonen layer and
// the Kohonen network

class Kohonen_network;


class Kohonen_layer: public layer
{

protected:

    float * weights;
    int winner_index;
    float win_distance;
    int neighborhood_size;

    friend Kohonen_network;

public:

    Kohonen_layer(int, int, int);
    ~Kohonen_layer();
    virtual void calc_out();
    void randomize_weights();
    void update_neigh_size(int);
    void update_weights(const float);
    void list_weights();
    void list_outputs();
    float get_win_dist();


};


class Kohonen_network

{

private:

    layer *layer_ptr[2];
    int layer_size[2];
    int neighborhood_size;

public:
    Kohonen_network();
    ~Kohonen_network();
    void get_layer_info();
    void set_up_network(int);
    void randomize_weights();
    void update_neigh_size(int);
    void update_weights(const float);
```

```
        void list_weights();
        void list_outputs();
        void get_next_vector(FILE *);
        void process_next_pattern();
        float get_win_dist();
        int get_win_index();

};
```

The **Kohonen_layer** is derived from the **layer** class, so it has pointers inherited that point to a set of outputs and a set of inputs. Let's look at some of the functions and member variables.

Kohonen_layer:

- **float * weights**   Pointer to the weights matrix

- **int winner_index**   Index value of the output which is the winner

- **float win_distance**   The Euclidean distance of the winner weight vector from the input vector

- **int neighborhood_size**   The size of the neighborhood

- **Kohonen_layer(int, int, int)**   Constructor for the layer: inputs, outputs, and the neighborhood size

- **~Kohonen_layer()**   destructor

- **virtual void calc_out()**   The function to calculate the outputs; for the Kohonen layer this models lateral competition

- **void randomize_weights()**   A function to initialize weights with random *normal* values

- **•void update_neigh_size(nt)**   This function updates the neighborhood size with a new value

- **void update_weights(const float)**   This function updates the weights according to the training law using the passed parameter, alpha

- **void list_weights()**   This function can be used to list the weight matrix

- **void list_outputs()**   This function is used to write outputs to the output file

- **float get_win_dist()**   Returns the Euclidean distance between the winner weight vector and the input vector.

Kohonen_network:

- ◆ **layer *layer_ptr[2]**   Pointer array; element 0 points to the input layer, element 1 points to the Kohonen layer
- ◆ **int layer_size[2]**   Array of layer sizes for the two layers
- ◆ **int neighborhood_size**   The current neighborhood size
- ◆ **Kohonen_network()**   Constructor
- ◆ **~Kohonen_network()**   Destructor
- ◆ **void get_layer_info()**   Gets information about the layer sizes
- ◆ **void set_up_network(int)**   Connects layers and sets up the Kohonen map
- ◆ **void randomize_weights()**   Creates random normalized weights
- ◆ **void update_neigh_size(int)**   Changes the neighborhood size
- ◆ **void update_weights(const float)**   Performs weight update according to the training law
- ◆ **void list_weights()**   Can be used to list the weight matrix
- ◆ **void list_outputs()**   Can be used to list outputs
- ◆ **void get_next_vector(FILE *)**   Function gets another input vector from the input file
- ◆ **void process_next_pattern()**   Applies pattern to the Kohonen map
- ◆ **float get_win_dist()**   Returns the winner's distance from the input vector
- ◆ **int get_win_index()**   Returns the index of the winner

Implementation of the Kohonen Layer and Kohonen Network

Listing 11.3 shows the *layerk.cpp* file, which has the implementation of the functions outlined.

Listing 11.3   Implementation file for the Kohonen layer and Kohonen network :layerk.cpp

```
// layerk.cpp    V.Rao, H.Rao
```

```
// compile for floating point hardware if available
#include "layer.cpp"
#include "layerk.h"

// -----------------------------------------
//            Kohonen layer
//-----------------------------------------


Kohonen_layer::Kohonen_layer(int i, int o, int
    init_neigh_size)
{
num_inputs=i;
num_outputs=o;
neighborhood_size=init_neigh_size;
weights = new float[num_inputs*num_outputs];
outputs = new float[num_outputs];
}


Kohonen_layer::~Kohonen_layer()
{
delete [num_outputs*num_inputs] weights;
delete [num_outputs] outputs;
}

void Kohonen_layer::calc_out()
{
// implement lateral competition
// choose the output with the largest
// value as the winner; neighboring
// outputs participate in next weight
// update. Winner's output is 1 while
// all other outputs are zero

int i,j,k;
float accumulator=0.0;
float maxval;
winner_index=0;
maxval=-1000000;

for (j=0; j<num_outputs; j++)
    {

    for (i=0; i<num_inputs; i++)

        {
        k=i*num_outputs;
        if (weights[k+j]*weights[k+j] > 1000000.0)
            {
```

```
            cout << "weights are blowing up\n";
            cout << "try a smaller learning constant\n";
            cout << "e.g. beta=0.02    aborting...\n";
            exit(1);
            }
        outputs[j]=weights[k+j]*(*(inputs+i));
        accumulator+=outputs[j];
        }
    // no squash function
    outputs[j]=accumulator;
    if (outputs[j] > maxval)
        {
        maxval=outputs[j];
        winner_index=j;
        }
    accumulator=0;
    }

// set winner output to 1
outputs[winner_index]=1.0;
// now zero out all other outputs
for (j=0; j< winner_index; j++)
    outputs[j]=0;
for (j=num_outputs-1; j>winner_index; j--)
    outputs[j]=0;


}


void Kohonen_layer::randomize_weights()
{
int i, j, k;
const unsigned first_time=1;

const unsigned not_first_time=0;
float discard;
float norm;

discard=randomweight(first_time);

for (i=0; i< num_inputs; i++)
    {
    k=i*num_outputs;
    for (j=0; j< num_outputs; j++)
        {
        weights[k+j]=randomweight(not_first_time);
        }
    }
```

```
// now need to normalize the weight vectors
// to unit length
// a weight vector is the set of weights for
// a given output

for (j=0; j< num_outputs; j++)
    {
    norm=0;
      for (i=0; i< num_inputs; i++)
        {
        k=i*num_outputs;
        norm+=weights[k+j]*weights[k+j];
        }
}
    norm = 1/((float)sqrt((double)norm));

for (i=0; i< num_inputs; i++)
        {
        k=i*num_outputs;
        weights[k+j]*=norm;
        }
    }


}


void Kohonen_layer::update_neigh_size(int new_neigh_size)
{
neighborhood_size=new_neigh_size;
}


void Kohonen_layer::update_weights(const float alpha)
{
int i, j, k;
int start_index, stop_index;
// learning law: weight_change =
//     alpha*(input-weight)
// zero change if input and weight
// vectors are aligned
// only update those outputs that
// are within a neighborhood's distance
// from the last winner
start_index = winner_index -
          neighborhood_size;


if (start_index < 0)
    start_index =0;
```

```
stop_index = winner_index +
        neighborhood_size;


if (stop_index > num_outputs-1)
    stop_index = num_outputs-1;



for (i=0; i< num_inputs; i++)
    {
    k=i*num_outputs;
    for (j=start_index; j<=stop_index; j++)
        weights[k+j] +=
            alpha*((*(inputs+i))-weights[k+j]);
    }

}

void Kohonen_layer::list_weights()
{
int i, j, k;

for (i=0; i< num_inputs; i++)
    {
    k=i*num_outputs;
    for (j=0; j< num_outputs; j++)
        cout << "weight["<<i<<","<<
            j<<"] is: "<<weights[k+j];
    }

}

void Kohonen_layer::list_outputs()
{
int i;

for (i=0; i< num_outputs; i++)
    {
    cout << "outputs["<<i<<
        "] is: "<<outputs[i];
    }

}


float Kohonen_layer::get_win_dist()
{
int i, j, k;
j=winner_index;
float accumulator=0;
```

```
float * win_dist_vec = new float [num_inputs];

for (i=0; i< num_inputs; i++)
    {
    k=i*num_outputs;
    win_dist_vec[i]=(*(inputs+i))-weights[k+j];
     accumulator+=win_dist_vec[i]*win_dist_vec[i];
     }

win_distance =(float)sqrt((double)accumulator);

delete [num_inputs]win_dist_vec;

return win_distance;

}


Kohonen_network::Kohonen_network()
{

}

Kohonen_network::~Kohonen_network()
{
}

void Kohonen_network::get_layer_info()
{
int i;

//-----------------------------------------
//
// Get layer sizes for the Kohonen network
//
// -----------------------------------------


cout << " Enter in the layer sizes separated by spaces.\n";
cout << " A Kohonen network has an input layer \n";
cout << " followed by a Kohonen (output) layer \n";

for (i=0; i<2; i++)
    {
    cin >> layer_size[i];
    }

// -------------------------------------------------------
```

```
// size of layers:
//    input_layer      layer_size[0]
//    Kohonen_layer    layer_size[1]
//------------------------------------------------------

}

void Kohonen_network::set_up_network(int nsz)
{
int i;

// set up neighborhood size
neighborhood_size = nsz;

//------------------------------------------------------
// Construct the layers
//
//------------------------------------------------------


layer_ptr[0] = new input_layer(0,layer_size[0]);

layer_ptr[1] =
    new Kohonen_layer(layer_size[0],
        layer_size[1],neighborhood_size);


for (i=0;i<2;i++)
    {
    if (layer_ptr[i] == 0)
        {
        cout << "insufficient memory\n";
        cout << "use a smaller architecture\n";
        exit(1);
        }
    }

//------------------------------------------------------
// Connect the layers
//
//------------------------------------------------------
// set inputs to previous layer outputs for the Kohonen layer

layer_ptr[1]->inputs = layer_ptr[0]->outputs;


}
```

```
void Kohonen_network::randomize_weights()
{

((Kohonen_layer *)layer_ptr[1])
     ->randomize_weights();
}


void Kohonen_network::update_neigh_size(int n)
{
((Kohonen_layer *)layer_ptr[1])
     ->update_neigh_size(n);
}

void Kohonen_network::update_weights(const float a)
{
((Kohonen_layer *)layer_ptr[1])
     ->update_weights(a);
}


void Kohonen_network::list_weights()
{
((Kohonen_layer *)layer_ptr[1])
     ->list_weights();
}

void Kohonen_network::list_outputs()
{
((Kohonen_layer *)layer_ptr[1])
     ->list_outputs();
}

void Kohonen_network::get_next_vector(FILE * ifile)
{
int i;
float normlength=0;
int num_inputs=layer_ptr[1]->num_inputs;
float *in = layer_ptr[1]->inputs;
// get a vector and normalize it
for (i=0; i<num_inputs; i++)
     {
     fscanf(ifile,"%f",(in+i));
     normlength += (*(in+i))*(*(in+i));
     }
fscanf(ifile,"\n");
normlength = 1/(float)sqrt((double)normlength);
for (i=0; i< num_inputs; i++)
     {
     (*(in+i)) *= normlength;
```

```
    }
}


void Kohonen_network::process_next_pattern()
{
    layer_ptr[1]->calc_out();
}

float Kohonen_network::get_win_dist()
{
float  retval;
retval=((Kohonen_layer *)layer_ptr[1])
        ->get_win_dist();

return retval;
}

int Kohonen_network::get_win_index()
{
return ((Kohonen_layer *)layer_ptr[1])
        ->winner_index;


}
```

Flow of the Program and the main() Function

The **main()** function is contained in a file called *kohonen.cpp*, which is shown in Listing 11.4. To compile this program, you need only compile and make this main file, kohonen.cpp. Other files are included in this.

Listing 11.4   The main implementation file, kohonen.cpp for the Kohonen Map program

```
// kohonen.cpp        V. Rao, H. Rao
// Program to simulate a Kohonen map

#include "layerk.cpp"

#define INPUT_FILE "input.dat"
#define OUTPUT_FILE "kohonen.dat"
#define dist_tol 0.05

void main()
{

int neighborhood_size, period;
```

```
float avg_dist_per_cycle=0.0;
float dist_last_cycle=0.0;
float avg_dist_per_pattern=100.0; // for the latest cycle
float dist_last_pattern=0.0;
float total_dist;
float alpha;
unsigned startup;
int max_cycles;
int patterns_per_cycle=0;

int total_cycles, total_patterns;



// create a network object
Kohonen_network knet;


FILE * input_file_ptr, * output_file_ptr;

// open input file for reading
if ((input_file_ptr=fopen(INPUT_FILE,"r"))==NULL)
        {
        cout << "problem opening input file\n";
        exit(1);
        }

// open writing file for writing
if ((output_file_ptr=fopen(OUTPUT_FILE,"w"))==NULL)
        {
        cout << "problem opening output file\n";
        exit(1);
        }



// -----------------------------------------
// Read in an initial values for alpha, and the
//  neighborhood size.
//  Both of these parameters are decreased with
//  time. The number of cycles to execute before
//  decreasing the value of these parameters is
//     called the period. Read in a value for the
//     period.
// -----------------------------------------
        cout << " Please enter initial values for:\n";
        cout << "alpha (0.01-1.0),\n";
        cout << "and the neighborhood size (integer between 0
        and
        50)\n";
        cout << "separated by spaces, e.g. 0.3 5 \n ";
```

```
        cin >> alpha >> neighborhood_size ;

        cout << "\nNow enter the period, which is the\n";
        cout << "number of cycles after which the values\n";
        cout << "for alpha the neighborhood size are
             decremented\n";
        cout << "choose an integer between 1 and 500 , e.g. 50
             \n";

        cin >> period;


        // Read in the maximum number of cycles
        // each pass through the input data file is a cycle
        cout << "\nPlease enter the maximum cycles for the
             simulation\n";
        cout << "A cycle is one pass through the data set.\n";
        cout << "Try a value of 500 to start with\n\n";

        cin >> max_cycles;


// the main loop
//
//      continue looping until the average distance is less
//      than the tolerance specified at the top of this file
//      , or the maximum number of
//      cycles is exceeded;

// initialize counters
total_cycles=0; // a cycle is once through all the input data
total_patterns=0; // a pattern is one entry in the input data



// get layer information
knet.get_layer_info();

// set up the network connections
knet.set_up_network(neighborhood_size);

// initialize the weights

// randomize weights for the Kohonen layer
// note that the randomize function for the
// Kohonen simulator generates
// weights that are normalized to length = 1
knet.randomize_weights();

// write header to output file
```

```
fprintf(output_file_ptr,
   "cycle\tpattern\twin index\tneigh_size\tavg_dist_per_pa
      tern\n");

fprintf(output_file_ptr,
   "---------------------------------------------------------
\n");

// main loop

startup=1;
total_dist=0;

while (
         (avg_dist_per_pattern > dist_tol)
         && (total_cycles < max_cycles)

         || (startup==1)
         )
{
startup=0;
dist_last_cycle=0; // reset for each cycle
patterns_per_cycle=0;
// process all the vectors in the datafile

while (!feof(input_file_ptr))
   {
   knet.get_next_vector(input_file_ptr);

   // now apply it to the Kohonen network
   knet.process_next_pattern();

  dist_last_pattern=knet.get_win_dist();

   // print result to output file
   fprintf(output_file_ptr,"%i\t%i\t%i\t\t%i\t\t%f\n",
    total_cycles,total_patterns,knet.get_win_index(),
    neighborhood_size,avg_dist_per_pattern);

   total_patterns++;

   // gradually reduce the neighborhood size
   // and the gain, alpha
   if (((total_cycles+1) % period) == 0)
      {
      if (neighborhood_size > 0)
         neighborhood_size --;
      knet.update_neigh_size(neighborhood_size);
      if (alpha>0.1)
         alpha -= (float)0.1;
      }
```

```
            patterns_per_cycle++;
            dist_last_cycle += dist_last_pattern;
            knet.update_weights(alpha);
            dist_last_pattern = 0;
        }

    avg_dist_per_pattern= dist_last_cycle/patterns_per_cycle;
    total_dist += dist_last_cycle;
    total_cycles++;



    fseek(input_file_ptr, 0L, SEEK_SET); // reset the file
                pointer
                // to the beginning of
                // the file


} // end main loop

cout << "\n\n\n\n\n\n\n\n\n\n\n";
cout << "------------------------------------------------
\n";
cout << " done \n";

avg_dist_per_cycle= total_dist/total_cycles;



cout << "\n";
cout << "---->average dist per cycle = " <<
avg_dist_per_cycle
    << " <---\n";
cout << "---->dist last cycle = " << dist_last_cycle << " <--
-
    \n";
cout << "->dist last cycle per pattern= " <<
    avg_dist_per_pattern << " <---\n";
cout << "----------->total cycles = " << total_cycles << "
<---\n";
cout << "----------->total patterns = " << total_patterns <<
" <---
    \n";
cout << "------------------------------------------------
\n";
// close the input file
fclose(input_file_ptr);
}
```

Flow of the Program

The flow of the program is very similar to the backpropagation simulator. The criterion for ending the simulation in the Kohonen program is the *average winner distance*. This is a Euclidean distance measure between the input vector and the winner's weight vector. This distance is the square root of the sum of the squares of the differences between individual vector components between the two vectors.

Results from Running the Kohonen Program

A Simple First Example

Once you compile the program, you need to create an input file to try it. We will first use a very simple input file and examine the results. Let us create an input file, *input.dat*, which contains only two arbitrary vectors:

```
0.4 0.98 0.1 0.2
0.5 0.22 0.8 0.9
```

NOTE

The file contains two four-dimensional vectors. We expect to see output that contains a different winner neuron for each of these patterns. If this is the case, then the Kohonen map has assigned different categories for each of the input vectors, and , in the future, you can expect to get the same winner classification for vectors that are close to or equal to these vectors.

By running the Kohonen map program you will see the following output (computer output is boldface):

Please enter initial values for:
alpha (0.01-1.0),
and the neighborhood size (integer between 0 and 50)
separated by spaces, e.g. 0.3 5
`0.3 5`

Now enter the period, which is the
number of cycles after which the values
for alpha the neighborhood size are decremented

choose an integer between I and 500 , e.g. 50
50

Please enter the maximum cycles for the simulation
A cycle is one pass through the data set.
Try a value of 500 to start with
500

Enter in the layer sizes separated by spaces.
A Kohonen network has an input layer
followed by a Kohonen (output) layer
4 10


```
-----------------------------------------------
      done

----->average dist per cycle = 0.544275 <-----
----->dist last cycle = 0.0827523 <-----
->dist last cycle per pattern= 0.0413762 <-----
------------>total cycles = I I <-----
------------>total patterns = 22 <-----
-----------------------------------------------
```

The layer sizes are given as 4 for the input layer and 10 for the Kohonen layer. You should choose the size of the Kohonen layer to be larger than the number of distinct patterns that you think are in the input data set. One of the outputs reported on the screen is the distance for the last cycle per pattern. This value is listed as 0.04, which is less than the terminating value set at the top of the kohonen.cpp file of 0.05. The map converged on a solution. Let us look at the file, *kohonen.dat*, the output file, to see the mapping to winner indexes.

```
cycle   pattern   win index   neigh_size   avg_dist_per_pattern
-------------------------------------------------------------------
0         0          1            5          100.000000
0         1          3            5          100.000000
1         2          1            5            0.304285
1         3          3            5            0.304285
2         4          1            5            0.568255
2         5          3            5            0.568255
3         6          1            5            0.542793
3         7          8            5            0.542793
4         8          1            5            0.502416
4         9          8            5            0.502416
5        10          1            5            0.351692
5        11          8            5            0.351692
6        12          1            5            0.246184
```

6	13	8	5	0.246184
7	14	1	5	0.172329
7	15	8	5	0.172329
8	16	1	5	0.120630
8	17	8	5	0.120630
9	18	1	5	0.084441
9	19	8	5	0.084441
10	20	1	5	0.059109
10	21	8	5	0.059109

In this example, the neighborhood size stays at its initial value of 5. In the first column you see the cycle number, and in the second the pattern number. Since there are two patterns per cycle, you see the cycle number repeated twice for each cycle.

The Kohonen map was able to find two distinct winner neurons for each of the patterns. One has winner index 1 and the other index 8.

N O T E

Orthogonal Input Vectors Example

For a second example, look at Figure 11.5, where we choose input vectors on a two-dimensional unit circle that are 90 degrees apart. The input.dat file should look like the following:

```
 1   0
 0   1
-1   0
 0  -1
```

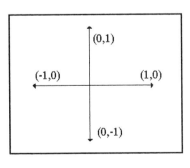

FIGURE 11.5 *Orthogonal input vectors*

Using the same parameters for the Kohonen network, but with layer sizes of 2 and 10, what result would you expect? Shown below is the output file, kohonen.dat.

cycle	pattern	win index	neigh_size	avg_dist_per_pattern
0	0	4	5	100.000000
0	1	0	5	100.000000
0	2	9	5	100.000000
0	3	3	5	100.000000
1	4	4	5	0.444558
1	5	0	5	0.444558
497	1991	6	0	0.707107
498	1992	0	0	0.707107
498	1993	0	0	0.707107
498	1994	6	0	0.707107
498	1995	6	0	0.707107
499	1996	0	0	0.707107
499	1997	0	0	0.707107
499	1998	6	0	0.707107
499	1999	6	0	0.707107

You can see that this example doesn't quite work. Even though the neighborhood size gradually got reduced to zero, the four inputs did not get categorized to different outputs. The winner distance became stuck at the value of 0.707, which is the distance from a vector at 45 degrees. In other words, the map generalizes a little too much arriving at the middle value for all of the input vectors.

You can fix this problem by starting with a smaller neighborhood size. This provides for less generalization. By using the same parameters and a neighborhood size of 2, the following output is obtained.

cycle	pattern	win index	neigh_size	avg_dist_per_pattern
0	0	5	2	100.000000
0	1	6	2	100.000000
0	2	4	2	100.000000
0	3	9	2	100.000000
1	4	0	2	0.431695
1	5	6	2	0.431695
1	6	3	2	0.431695
1	7	9	2	0.431695
2	8	0	2	0.504728
2	9	6	2	0.504728
2	10	3	2	0.504728
2	11	9	2	0.504728

3	12	0	2	0.353309
3	13	6	2	0.353309
3	14	3	2	0.353309
3	15	9	2	0.353309
4	16	0	2	0.247317
4	17	6	2	0.247317
4	18	3	2	0.247317
4	19	9	2	0.247317
5	20	0	2	0.173122
5	21	6	2	0.173122
5	22	3	2	0.173122
5	23	9	2	0.173122
6	24	0	2	0.121185
6	25	6	2	0.121185
6	26	3	2	0.121185
6	27	9	2	0.121185
7	28	0	2	0.084830
7	29	6	2	0.084830
7	30	3	2	0.084830
7	31	9	2	0.084830
8	32	0	2	0.059381
8	33	6	2	0.059381
8	34	3	2	0.059381
8	35	9	2	0.059381

For this case, the network quickly converges on a unique winner for each of the four input patterns, and the distance criterion is below the set criterion within eight cycles. You can experiment with other input data sets and combinations of Kohonen network parameters.

Other Variations of Kohonen Networks

There are many variations of the Kohonen network. Some of these will be briefly discussed in this section.

Using a Conscience

DeSieno has used a *conscience* factor in a Kohonen network. For a winning neuron, if the neuron is winning more than a fair share of the time, then this neuron has a threshold that is applied temporarily to allow other neurons the chance to win.

LVQ: Learning Vector Quantizer

You have heard about LVQ (Learning Vector Quantizer). In light of the Kohonen map, it should be pointed out that the LVQ is simply a supervised version of the Kohonen network. Inputs and expected output categories are presented to the network for training.

Counterpropagation Network

A neural network topology, called a *counterpropagation* network, is a combination of a Kohonen layer with a Grossberg layer. This network was developed by Robert Hecht-Nielsen and is useful for prototyping of systems, with a fairly rapid training time compared to backpropagation. The Kohonen layer provides for categorization, while the Grossberg layer allows for Hebbian conditioned learning. Counterpropagation has been used successfully in data compression applications.

Summary

- ◆ The Kohonen feature map is an example of an unsupervised neural network that is mainly used as a classifier system. As more inputs are presented to this network, the network improves its learning and is able to adapt to changing inputs.

- ◆ The training law for the Kohonen network tries to align the weight vectors along the same direction as input vectors.

- ◆ The Kohonen network models lateral competition as a form of self-organization. One winner neuron is derived for each input pattern to categorize that input.

- ◆ Only neurons within a certain distance from the winner are allowed to participate in training for a given input pattern.

Chapter

12

Application to Pattern Recognition

Using the Kohonen Feature Map

n this chapter, you will use the Kohonen program developed in Chapter 11 to recognize patterns. You will modify the Kohonen program for the display of patterns.

An Example Problem: Character Recognition

The problem that is presented in this chapter is to recognize or categorize alphabetic characters. You will input various alphabetic characters to a Kohonen map and train the network to recognize these as separate categories. This program can be used to try other experiments that will be discussed at the end of this chapter.

251

Representing Characters

Each character is represented by a 5 x 7 grid of pixels. We use the graphical printing characters of the ASCII character set to show a gray scale output for each pixel. To represent the letter A, for example, you could use the pattern shown in Figure 12.1. Here the blackened boxes represent value 1, while empty boxes represent a zero. You can represent all characters this way, with a binary map of 35 pixel values.

The letter A is represented by the values:

```
0 0 I 0 0
0 I 0 I 0
I 0 0 0 I
I 0 0 0 I
I I I I I
I 0 0 0 I
I 0 0 0 I
```

For use in the Kohonen program, we need to serialize the rows, so that all entries appear on one line.

For the characters A and X you would end up with the following entries in the input file, input.dat:

```
0 0 1 0 0   0 1 0 1 0  1 0 0 0 1  1 0 0 0 1  1 1 1 1 1  1 0 0 0 1  1 0 0 0 1
<< the letter A
1 0 0 0 1   0 1 0 1 0  0 0 1 0 0  0 0 1 0 0  0 0 1 0 0  0 1 0 1 0  1 0 0 0 1
<< the letter X
```

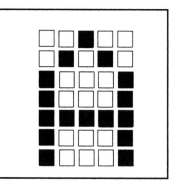

FIGURE 12.1 *Representation of the letter A with a 5 x 7 pattern.*

Monitoring the Weights

We will present the Kohonen map with many such characters and find the response in output. You will be able to watch the Kohonen map as it goes through its cycles and learns the input patterns. At the same time, you should be able to watch the weight vectors for the winner neurons to see the pattern that is developing in the weights. Remember that for a Kohonen map the weight vectors tend to become aligned with the input vectors. So after a while, you will notice that the weight vector for the input will resemble the input pattern that you are categorizing.

Representing the Weight Vector

Although on and off values are fine for the input vectors mentioned, you need to see gray scale values for the weight vector. This can be accomplished by quantizing the weight vector into four bins, each represented by a different ASCII graphic character:

$< =0$	white rectangle (space)
$0< \text{weight} < =0.25$	light-dotted rectangle
$0.25 < \text{weight} < =0.50$	medium-dotted rectangle
$0.50< \text{weight} <=0.75$	dark-dotted rectangle
$\text{weight} > 0.75$	black rectangle

The ASCII values for the graphics characters to be used are listed in Table 12.1.

TABLE 12.1 *ASCII values for rectangle graphic characters*

character	ASCII decimal value
white rectangle	255
light-dotted rectangle	176
medium-dotted rectangle	177
dark-dotted rectangle	178
black rectangle	219

C++ Code Development

The changes to the Kohonen program are relatively minor.

Changes to the Kohonen Program

The first change to make is to the **Kohonen_network** class definition. This is in the file, *layerk.h*, shown in Listing 12.1.

Listing 12.1   Updated layerk.h file

```
class Kohonen_network

{

private:

    layer *layer_ptr[2];
    int layer_size[2];
    int neighborhood_size;

public:
    Kohonen_network();
    ~Kohonen_network();
    void get_layer_info();
    void set_up_network(int);
    void randomize_weights();
    void update_neigh_size(int);
    void update_weights(const float);
    void list_weights();
    void list_outputs();
    void get_next_vector(FILE *);
    void process_next_pattern();
    float get_win_dist();
    int get_win_index();
    void display_input_char();
    void display_winner_weights();

    };
```

The new **member** functions are shown in boldface. The functions **display_input_char()** and **display_winner_weights()** are used to display the input and weight maps on the screen to watch weight character map converge to the input map.

The implementation of these functions is in the file, *layerk.cpp*. The portion of this file containing these functions is shown in Listing 12.2.

Listing 12.2   Additions to the layerk.cpp implementation file

```
void Kohonen_network::display_input_char()
{
int i, num_inputs;
unsigned char ch;
float temp;
int col=0;
float * inputptr;

num_inputs=layer_ptr[1]->num_inputs;
inputptr = layer_ptr[1]->inputs;
// we've got a 5x7 character to display

for (i=0; i<num_inputs; i++)
    {
    temp = *(inputptr);
    if (temp <= 0)
       ch=255;// blank
    else if ((temp > 0) && (temp <= 0.25))
        ch=176; // dotted rectangle -light
    else if ((temp > 0.25) && (temp <= 0.50))
        ch=177; // dotted rectangle -medium
    else if ((temp >0.50) && (temp <= 0.75))
        ch=178; // dotted rectangle -dark
    else if (temp > 0.75)
        ch=219; // filled rectangle
    printf("%c",ch); //fill a row
    col++;
    if ((col % 5)==0)
        printf("\n"); // new row
    inputptr++;
    }
printf("\n\n\n");
}

void Kohonen_network::display_winner_weights()
{
int i, k;
unsigned char ch;
float temp;
float * wmat;
int col=0;
int win_index;
int num_inputs, num_outputs;

num_inputs= layer_ptr[1]->num_inputs;
wmat = ((Kohonen_layer*)layer_ptr[1])
```

```
            ->weights;
win_index=((Kohonen_layer*)layer_ptr[1])
        ->winner_index;

num_outputs=layer_ptr[1]->num_outputs;

// we've got a 5x7 character to display

for (i=0; i<num_inputs; i++)
    {
    k= i*num_outputs;
    temp = wmat[k+win_index];
    if (temp <= 0)
        ch=255;// blank
    else if ((temp > 0) && (temp <= 0.25))
        ch=176; // dotted rectangle -light
    else if ((temp > 0.25) && (temp <= 0.50))
        ch=177; // dotted rectangle -medium
    else if ((temp > 0.50) && (temp <= 0.75))
        ch=178; // dotted rectangle -dark
    else if (temp > 0.75)
        ch=219; // filled rectangle
    printf("%c",ch); //fill a row
    col++;
    if ((col % 5)==0)
        printf("\n"); // new row
    }
printf("\n\n");
printf("-----------------\n");

}
```

The final change to make is to the kohonen.cpp file. The new file is called
pattern.cpp and is shown in Listing 12.3.

Listing 12.3   The implementation file pattern.cpp
```
// pattern.cpp       V. Rao, H. Rao
// Kohonen map for pattern recognition
#include "layerk.cpp"


#define INPUT_FILE "input.dat"
#define OUTPUT_FILE "kohonen.dat"
#define dist_tol 0.001
#define wait_cycles   10000 // creates a pause to
                // view the character maps


void main()
{
```

```
int neighborhood_size, period;
float avg_dist_per_cycle=0.0;
float dist_last_cycle=0.0;
float avg_dist_per_pattern=100.0; // for the latest cycle
float dist_last_pattern=0.0;
float total_dist;
float alpha;
unsigned startup;
int max_cycles;
int patterns_per_cycle=0;

int total_cycles, total_patterns;
int i;


// create a network object
Kohonen_network knet;


FILE * input_file_ptr, * output_file_ptr;

// open input file for reading
if ((input_file_ptr=fopen(INPUT_FILE,"r"))==NULL)
        {
        cout << "problem opening input file\n";
        exit(1);
        }

// open writing file for writing
if ((output_file_ptr=fopen(OUTPUT_FILE,"w"))==NULL)
        {
        cout << "problem opening output file\n";
        exit(1);
        }



// ---------------------------------------------
// Read in an initial values for alpha, and the
// neighborhood size.
// Both of these parameters are decreased with
// time. The number of cycles to execute before
// decreasing the value of these parameters is
//    called the period. Read in a value for the
//    period.
// ---------------------------------------------
        cout << " Please enter initial values for:\n";
        cout << "alpha (0.01-1.0),\n";
        cout << "and the neighborhood size (integer between 0\
            and 50)\n";
```

```
            cout << "separated by spaces, e.g. 0.3 5 \n ";

            cin >> alpha >> neighborhood_size ;

            cout << "\nNow enter the period, which is the\n";
            cout << "number of cycles after which the values\n";
            cout << "for alpha the neighborhood size are \
                decremented\n";
            cout << "choose an integer between 1 and 500 , e.g. \
                50 \n";

            cin >> period;


    // Read in the maximum number of cycles
    // each pass through the input data file is a cycle
            cout << "\nPlease enter the maximum cycles for the
            simulation\n";
            cout << "A cycle is one pass through the data set.\n";
            cout << "Try a value of 500 to start with\n\n";

            cin >> max_cycles;


// the main loop
//
//      continue looping until the average distance is less
than
//      the tolerance specified at the top of this file
//      , or the maximum number of
//      cycles is exceeded;

// initialize counters
total_cycles=0; // a cycle is once through all the input data
total_patterns=0; // a pattern is one entry in the input data



// get layer information
knet.get_layer_info();

// set up the network connections
knet.set_up_network(neighborhood_size);

// initialize the weights

// randomize weights for the Kohonen layer
// note that the randomize function for the
// Kohonen simulator generates
// weights that are normalized to length = 1
knet.randomize_weights();
```

```
// write header to output file
fprintf(output_file_ptr,
    "cycle\tpattern\twin index\tneigh_size\\
        tavg_dist_per_pattern\n");

fprintf(output_file_ptr,
    "---------------------------------------------\n");


startup=1;
total_dist=0;

while (
        (avg_dist_per_pattern > dist_tol)
        && (total_cycles < max_cycles)

        || (startup==1)
        )
{
startup=0;
dist_last_cycle=0; // reset for each cycle
patterns_per_cycle=0;
// process all the vectors in the datafile


while (!feof(input_file_ptr))
    {
    knet.get_next_vector(input_file_ptr);

    // now apply it to the Kohonen network
    knet.process_next_pattern();

    dist_last_pattern=knet.get_win_dist();

    // print result to output file
    fprintf(output_file_ptr,"%i\t%i\t%i\t\t%i\t\t%f\n",
      total_cycles,total_patterns,knet.get_win_index(),
      neighborhood_size,avg_dist_per_pattern);

    // display the input character and the
    // weights for the winner to see match

    knet.display_input_char();
    knet.display_winner_weights();
    // pause for a while to view the
    // character maps
    for (i=0; i<wait_cycles; i++)
```

```
            {;}

        total_patterns++;

        // gradually reduce the neighborhood size
        // and the gain, alpha
        if (((total_cycles+1) % period) == 0)
            {
            if (neighborhood_size > 0)
                neighborhood_size --;
            knet.update_neigh_size(neighborhood_size);
            if (alpha>0.1)
                alpha -= (float)0.1;
            }


        patterns_per_cycle++;
        dist_last_cycle += dist_last_pattern;
        knet.update_weights(alpha);
        dist_last_pattern = 0;
         }
    avg_dist_per_pattern= dist_last_cycle/patterns_per_cycle;
    total_dist += dist_last_cycle;
    total_cycles++;




    fseek(input_file_ptr, 0L, SEEK_SET); // reset the file
        pointer
                    // to the beginning of
                    // the file


    } // end main loop

    cout << "\n\n\n\n\n\n\n\n\n\n\n";
    cout << "-----------------------------------------------\n";
    cout << " done \n";

    avg_dist_per_cycle= total_dist/total_cycles;


    cout << "\n";
    cout << "---->average dist per cycle = " <<
    avg_dist_per_cycle << " <---\n";
    cout << ">dist last cycle = " << dist_last_cycle << " < \n";
    cout << "->dist last cycle per pattern= " <<
```

```
    avg_dist_per_pattern << " <---\n";
cout << "----->total cycles = " << total_cycles << " <---\n";
cout << "----------->total patterns = " <<
    total_patterns << " <---\n";
cout << "---------------------------------------------------\n";
// close the input file
fclose(input_file_ptr);
}
```

Changes to the program are indicated in boldface. Compile this program by compiling and making the pattern.cpp file, after modifying the layerk.cpp and layerk.h files as indicated previously.

Testing the Program

Let us run the example that we have created an input file for. We have an input.dat file with the characters A and X defined. A run of the program with these inputs is shown as follows:

Please enter initial values for:
alpha (0.01-1.0),
and the neighborhood size (integer between 0 and 50)
separated by spaces, e.g. 0.3 5
0.3 5
Now enter the period, which is the
number of cycles after which the values
for alpha the neighborhood size are decremented
choose an integer between 1 and 500 , e.g., 50
50
Please enter the maximum cycles for the simulation
A cycle is one pass through the data set.
Try a value of 500 to start with
500
Enter in the layer sizes separated by spaces.
A Kohonen network has an input layer
followed by a Kohonen (output) layer
35 100

The output of the program is contained in file kohonen.dat as usual. This shows the following result.

cycle	pattern	win index	neigh_size	avg_dist_per_pattern
0	0	42	5	100.000000
0	1	47	5	100.000000
1	2	42	5	0.508321

1	3	47	5	0.508321
2	4	40	5	0.742254
2	5	47	5	0.742254
3	6	40	5	0.560121
3	7	47	5	0.560121
4	8	40	5	0.392084
4	9	47	5	0.392084
5	10	40	5	0.274459
5	11	47	5	0.274459
6	12	40	5	0.192121
6	13	47	5	0.192121
7	14	40	5	0.134485
7	15	47	5	0.134485
8	16	40	5	0.094139
8	17	47	5	0.094139
9	18	40	5	0.065898
9	19	47	5	0.065898
10	20	40	5	0.046128
10	21	47	5	0.046128
11	22	40	5	0.032290
11	23	47	5	0.032290
12	24	40	5	0.022603
12	25	47	5	0.022603
13	26	40	5	0.015822
13	27	47	5	0.015822
14	28	40	5	0.011075
14	29	47	5	0.011075
15	30	40	5	0.007753
15	31	47	5	0.007753
16	32	40	5	0.005427
16	33	47	5	0.005427
17	34	40	5	0.003799
17	35	47	5	0.003799
18	36	40	5	0.002659
18	37	47	5	0.002659
19	38	40	5	0.001861
19	39	47	5	0.001861
20	40	40	5	0.001303
20	41	47	5	0.001303

The tolerance for the distance was set to be 0.001 for this program, and the program was able to converge to this value. Both of the inputs were successfully classified into two different winning output neurons. In Figures 12.2 and 12.3 you see two snapshots of the input and weight vectors that you will find with this program. The weight vector resembles the input as you can see, but is not an exact replication.

FIGURE 12.2   *Sample screen output of the letter A from the input and weight vectors.*

FIGURE 12.3   *Sample screen output of the letter X from the input and weight vectors.*

Generalization versus Memorization

As mentioned in Chapter 11, you actually don't desire the exact replication of the input pattern for the weight vector. This would amount to memorizing of the input patterns with no capacity for generalization.

For example, a typical use of this alphabet classifier system would be to use it to process noisy data, like handwritten characters. In such a case, you would need a great deal of latitude in scoping a class for a letter A.

Adding Characters

The next step of the program is to add characters and see what categories they end up in. There are many alphabetic characters that look alike, such as H and B for example. You can expect the Kohonen classifier to group these like characters into the same class.

We now modify the input.dat file to add the characters H, B, and I. The new input.dat file is shown as follows.

```
0 0 1 0 0   0 1 0 1 0   1 0 0 0 1   1 0 0 0 1   1 1 1 1 1   1 0 0 0 1   1 0 0 0 1
1 0 0 0 1   0 1 0 1 0   0 0 1 0 0   0 0 1 0 0   0 0 1 0 0   0 1 0 1 0   1 0 0 0 1
1 0 0 0 1   1 0 0 0 1   1 0 0 0 1   1 1 1 1 1   1 0 0 0 1   1 0 0 0 1   1 0 0 0 1
1 1 1 1 1   1 0 0 0 1   1 0 0 0 1   1 1 1 1 1   1 0 0 0 1   1 0 0 0 1   1 1 1 1 1
0 0 1 0 0   0 0 1 0 0   0 0 1 0 0   0 0 1 0 0   0 0 1 0 0   0 0 1 0 0   0 0 1 0 0
```

The output using this input file is shown as follows.

```
-------------------------------------------------
     done

---->average dist per cycle = 0.732607 <---
---->dist last cycle = 0.00360096 <---
->dist last cycle per pattern= 0.000720192 <---
----------->total cycles = 37 <---
----------->total patterns = 185 <---
-------------------------------------------------
```

The file kohonen.dat with the output values is now shown as follows.

cycle	pattern	win index	neigh_size	avg_dist_per_pattern
0	0	69	5	100.000000
0	1	93	5	100.000000
0	2	18	5	100.000000
0	3	18	5	100.000000
0	4	78	5	100.000000
1	5	69	5	0.806743
1	6	93	5	0.806743
1	7	18	5	0.806743
1	8	18	5	0.806743
1	9	78	5	0.806743
2	10	69	5	0.669678
2	11	93	5	0.669678
2	12	18	5	0.669678
2	13	18	5	0.669678
2	14	78	5	0.669678

3	15	69	5	0.469631
3	16	93	5	0.469631
3	17	18	5	0.469631
3	18	18	5	0.469631
3	19	78	5	0.469631
4	20	69	5	0.354791
4	21	93	5	0.354791
4	22	18	5	0.354791
4	23	18	5	0.354791
4	24	78	5	0.354791
5	25	69	5	0.282990
5	26	93	5	0.282990
5	27	18	5	0.282990
. . .				
35	179	78	5	0.001470
36	180	69	5	0.001029
36	181	93	5	0.001029
36	182	13	5	0.001029
36	183	19	5	0.001029
36	184	78	5	0.001029

Again, the network does not find a problem in classifying these vectors.

> Until cycle 21, both the H and the B were classified as output neuron 18. The ability to distinguish these vectors is largely due to the small tolerance we have assigned as a termination criterion.
>
> N O T E

Other Experiments to Try

You can try a number of other experiments with the program. For example, you can repeat the input file but with the order of the entries changed. In other words, you can present the same inputs a number of times in different order. This actually helps the Kohonen network train faster. You can try applying garbled versions of the characters to see if the network distinguishes them. Just as in the backpropagation program, you can save the weights in a weight file to freeze the state of training, and then apply new inputs. You can enter all of the characters from A to Z and see the classification that results. Do you need to train on all of the characters or a subset? You can change the size of the Kohonen layer. How many neurons do you need to recognize the complete alphabet?

There is no restriction on using digital inputs of 1 and 0 as we had used. You can apply gray scale analog values. The program will display the input pattern according to the quantization levels that were set. This set can be expanded, and you can use a graphics interface to display more levels. You can then try pattern recognition of arbitrary images, but remember that processing time will increase rapidly with the number of neurons used. The number of input neurons you choose is dictated by the image resolution, unless you *filter* and/or *subsample* the image before you present it to the network. Filtering is the process of using a type of averaging function applied to groups of pixels. Subsampling is the process of choosing a lower-output image resolution by selecting fewer pixels than a source image. If you start with an image that is 100 x 100 pixels, you can subsample this image 2:1 in each direction to obtain an image that is one-fourth the size, or 50 x 50 pixels. Whether you throw away every other pixel to get this output resolution or apply a filter is up to you. You could average every two pixels to get one output pixel as an example of a very simple filter.

Summary

◆ This chapter presented a simple character recognition program using a Kohonen feature map.

◆ The input vectors and the weight vectors were displayed to show convergence and note similarity between the two vectors.

◆ As training progresses, the weight vector for the winner neuron resembles the input character map.

Chapter

13

Backpropagation II

Enhancing the Simulator

n Chapter 7, you developed a backpropagation simulator. In this chapter, you will put it to use with examples and also add some new features to the simulator: a term called *momentum*, and the capability of adding noise to the inputs during simulation. There are many variations of the algorithm that try to alleviate two problems with backpropagation: First, like other neural networks, there is a strong possibility that the solution found with backpropagation is not a global error minimum, but a local one. You may need to shake the weights a little by some means to get out of the local minimum, and possibly arrive at a lower minimum. The second problem with backpropagation is speed. The algorithm is very slow at learning. There are many proposals for speeding up the search process. Neural networks are inherently parallel processing architectures and are suited for simulation on parallel processing hardware. While there are a few plug-in neural net or digital signal processing

267

boards available in the market, the low-cost simulation platform of choice remains the personal computer. Speed enhancements to the training algorithm are therefore very necessary.

Another Example of Using Backpropagation

Before modifying the simulator to add features, let's look at the same problem we used the Kohonen map to analyze in Chapter 12. As you recall, we would like to be able to distinguish alphabetic characters by assigning them to different bins. For backpropagation, we would apply the inputs and train the network with anticipated responses. Here is the input file that we used for distinguishing five different characters, A, X, H, B, and I:

```
0 0 1 0 0   0 1 0 1 0   1 0 0 0 1   1 0 0 0 1   1 1 1 1 1   1 0 0 0 1   1 0 0 0 1
1 0 0 0 1   0 1 0 1 0   0 0 1 0 0   0 0 1 0 0   0 0 1 0 0   0 1 0 1 0   1 0 0 0 1
1 0 0 0 1   1 0 0 0 1   1 0 0 0 1   1 1 1 1 1   1 0 0 0 1   1 0 0 0 1   1 0 0 0 1
1 1 1 1 1   1 0 0 0 1   1 0 0 0 1   1 1 1 1 1   1 0 0 0 1   1 0 0 0 1   1 1 1 1 1
0 0 1 0 0   0 0 1 0 0   0 0 1 0 0   0 0 1 0 0   0 0 1 0 0   0 0 1 0 0   0 0 1 0 0
```

Each line has a 5 x 7 dot representation of each character. Now we need to name each of the output categories. We can assign a simple 3-bit representation as follows:

```
A   000
X   010
H   100
B   101
I   111
```

Let's train the network to recognize these characters. The training.dat file looks like the following:

```
0 0 1 0 0  0 1 0 1 0  1 0 0 0 1  1 0 0 0 1  1 1 1 1 1  1 0 0 0 1  1 0 0 0 1  0 0 0
1 0 0 0 1  0 1 0 1 0  0 0 1 0 0  0 0 1 0 0  0 0 1 0 0  0 1 0 1 0  1 0 0 0 1  0 1 0
1 0 0 0 1  1 0 0 0 1  1 0 0 0 1  1 1 1 1 1  1 0 0 0 1  1 0 0 0 1  1 0 0 0 1  1 0 0
1 1 1 1 1  1 0 0 0 1  1 0 0 0 1  1 1 1 1 1  1 0 0 0 1  1 0 0 0 1  1 1 1 1 1  1 0 1
0 0 1 0 0  0 0 1 0 0  0 0 1 0 0  0 0 1 0 0  0 0 1 0 0  0 0 1 0 0  0 0 1 0 0  1 1 1
```

You can now start the simulator. Using the parameters (beta = 0.1, tolerance = 0.001, and max_cycles = 1000) and with three layers of size 35 (input), 5 (middle), and 3 (output), you will get a typical result like the following:

```
-----------------------------------------------------
   done:  results in file output.dat
          training: last vector only
          not training: full cycle

          weights saved in file weights.dat

---->average error per cycle = 0.035713<---
---->error last cycle = 0.008223 <---
->error last cycle per pattern= 0.00164455 <---
----------->total cycles = 1000 <---
----------->total patterns = 5000 <---
-----------------------------------------------------
```

The simulator stopped at the 1,000 maximum cycles specified in this case. Your results will be different since the weights start at a random point. Note that the tolerance specified was nearly met. Let us see how close the output came to what we wanted. Look at the output.dat file. You can see the match for the last pattern as follows:

```
for input vector:
0.000000  0.000000  1.000000  0.000000  0.000000  0.000000
   0.000000  1.000000  0.000000  0.000000  0.000000  0.000000
   1.000000  0.000000  0.000000  0.000000  0.000000  1.000000
   0.000000  0.000000  0.000000  0.000000  1.000000  0.000000
   0.000000  0.000000  0.000000  1.000000  0.000000  0.000000
   0.000000  0.000000  1.000000  0.000000  0.000000
output vector is:
0.999637  0.998721  0.999330
expected output vector is:
1.000000  1.000000  1.000000
---------------------
```

To see the outputs of all the patterns, we need to copy the training.dat file to the test.dat file and rerun the simulator in **Test** mode. Remember to delete the expected output field once you copy the file.

Running the simulator in **Test** mode (0) shows the following result in the output.dat file:

```
for input vector:
0.000000  0.000000  1.000000  0.000000  0.000000  0.000000
   1.000000  0.000000  1.000000  0.000000  1.000000  0.000000
   0.000000  0.000000  1.000000  1.000000  0.000000  0.000000
   0.000000  1.000000  1.000000  1.000000  1.000000  1.000000
   1.000000  1.000000  0.000000  0.000000  0.000000  1.000000
   1.000000  0.000000  0.000000  0.000000  1.000000
output vector is:
```

```
0.005010   0.002405   0.000141
---------------------
for input vector:
1.000000   0.000000   0.000000   0.000000   1.000000   0.000000
    1.000000   0.000000   1.000000   0.000000   0.000000   0.000000
    1.000000   0.000000   0.000000   0.000000   0.000000   1.000000
    0.000000   0.000000   0.000000   0.000000   1.000000   0.000000
    0.000000   0.000000   1.000000   0.000000   1.000000   0.000000
    1.000000   0.000000   0.000000   0.000000   1.000000
output vector is:
0.001230   0.997844   0.000663
---------------------
for input vector:
1.000000   0.000000   0.000000   0.000000   1.000000   1.000000
    0.000000   0.000000   0.000000   1.000000   1.000000   0.000000
    0.000000   0.000000   1.000000   1.000000   1.000000   1.000000
    1.000000   1.000000   1.000000   0.000000   0.000000   0.000000
    1.000000   1.000000   0.000000   0.000000   0.000000   1.000000
    1.000000   0.000000   0.000000   0.000000   1.000000
output vector is:
0.995348   0.000253   0.002677
---------------------
for input vector:
1.000000   1.000000   1.000000   1.000000   1.000000   1.000000
    0.000000   0.000000   0.000000   1.000000   1.000000   0.000000
    0.000000   0.000000   1.000000   1.000000   1.000000   1.000000
    1.000000   1.000000   1.000000   0.000000   0.000000   0.000000
    1.000000   1.000000   0.000000   0.000000   0.000000   1.000000
    1.000000   1.000000   1.000000   1.000000   1.000000
output vector is:
0.999966   0.000982   0.997594
---------------------
for input vector:
0.000000   0.000000   1.000000   0.000000   0.000000   0.000000
    0.000000   1.000000   0.000000   0.000000   0.000000   0.000000
    1.000000   0.000000   0.000000   0.000000   0.000000   1.000000
    0.000000   0.000000   0.000000   0.000000   1.000000   0.000000
    0.000000   0.000000   0.000000   1.000000   0.000000   0.000000
    0.000000   0.000000   1.000000   0.000000   0.000000
output vector is:
0.999637   0.998721   0.999330
---------------------
```

The training patterns are learned very well. If a smaller tolerance is used, it would be possible to complete the learning in fewer cycles. What happens if we present a foreign character to the network? Let us create a new test.dat file with two entries for the letters M and J, as follows:

```
1 0 0 0 1   1 1 0 1 1   1 0 1 0 1   1 0 0 0 1   1 0 0 0 1   1 0 0 0 1   1 0 0 0 1
```

```
00100  00100  00100  00100  00100  00100  01111
```

The results should show each foreign character in the category closest to it. The middle layer of the network acts as a feature detector. Since we specified five neurons, we have given the network the freedom to define five features in the input training set to use to categorize inputs. The results in the output.dat file are shown as follows:

```
for input vector:
1.000000  0.000000  0.000000  0.000000  1.000000  1.000000
    1.000000  0.000000  1.000000  1.000000  1.000000  0.000000
    1.000000  0.000000  1.000000  1.000000  0.000000  0.000000
    0.000000  1.000000  1.000000  0.000000  0.000000  0.000000
    1.000000  1.000000  0.000000  0.000000  0.000000  1.000000
    1.000000  0.000000  0.000000  0.000000  1.000000
output vector is:
0.963513  0.000800  0.001231
- - - - - - - - - - - - - - - - - - - - -
for input vector:
0.000000  0.000000  1.000000  0.000000  0.000000  0.000000
    0.000000  1.000000  0.000000  0.000000  0.000000  0.000000
    1.000000  0.000000  0.000000  0.000000  0.000000  1.000000
    0.000000  0.000000  0.000000  0.000000  1.000000  0.000000
    0.000000  0.000000  0.000000  1.000000  0.000000  0.000000
    0.000000  1.000000  1.000000  1.000000  1.000000
output vector is:
0.999469  0.996339  0.999157
- - - - - - - - - - - - - - - - - - - - -
```

In the first pattern, an M is categorized as an H, whereas in the second pattern, a J is categorized as an I as expected. The case of the first pattern seems reasonable since the H and M share many pixels in common.

Other Experiments to Try

There are many other experiments you could try, in order to get a better feel for how to train and use a backpropagation neural network.

- ◆ You could use the ASCII 8-bit code to represent each character, and try to train the network. You could also code all of the alphabetic characters and see if it's possible to distinguish all of them.
- ◆ You can garble a character, to see if you still get the correct output.
- ◆ You could try changing the size of the middle layer, and see the effect on training time and generalization ability.

◆ You could change the tolerance setting to see the difference between an overtrained and undertrained network in generalization capability. That is, given a foreign pattern, is the network able to find the closest match and use that particular category, or does it arrive at a new category altogether?

We will return to the same example after enhancing the simulator with momentum and noise addition capability.

Adding the Momentum Term

A simple change to the training law that sometimes results in much faster training is the addition of a *momentum term*. The training law for backpropagation as implemented in the simulator is:

```
Weight change = Beta * output_error * input
```

Now we add a term to the weight change equation as follows:

```
Weight change = Beta * output_error * input +
    Alpha*previous_weight_change
```

The second term in this equation is the momentum term. The weight change, in the absence of error, would be a constant multiple by the previous weight change. In other words, the weight change continues in the direction it was heading. The momentum term is an attempt to try to keep the weight change process moving, and thereby not get stuck in local minimas.

Code Changes

The affected files to implement this change are the layer.cpp file , to modify the **update_weights() Member** function of the **output_layer** class, and the main backprop.cpp file to read in the value for alpha and pass it to the **Member** function. There is some additional storage needed for storing previous weight changes, and this affects the layer.h file. The momentum term could be implemented in two ways:

1. Using the weight change for the previous pattern.

2. Using the weight change accumulated over the previous cycle.

Although both of these implementations are valid, the second is particularly useful, since it adds a term that is significant for all patterns, and hence would contribute to global error reduction. We implement the second choice by accumulating the value of the current cycle weight changes in a vector called **cum_deltas**. The past cycle weight changes are stored in a vector called **past_deltas**. These are shown as follows in a portion of the layer.h file.

```
class output_layer: public layer
{
protected:

    float * weights;
    float * output_errors; // array of errors at output
    float * back_errors; // array of errors back-propagated
    float * expected_values;   // to inputs
    float * cum_deltas; // for momentum
    float * past_deltas;// for momentum

    friend network;
    ...
```

Changes to the layer.cpp File

The implementation file for the **Layer** class changes in the **output_layer::update_weights()** routine and the constructor and destructor for **output_layer**. First, here is the constructor for **output_layer**. Changes are highlighted in boldface.

```
output_layer::output_layer(int ins, int outs)
{
int i, j, k;
num_inputs=ins;
num_outputs=outs;
weights = new float[num_inputs*num_outputs];
output_errors = new float[num_outputs];
back_errors = new float[num_inputs];
outputs = new float[num_outputs];
expected_values = new float[num_outputs];
cum_deltas = new float[num_inputs*num_outputs];
past_deltas = new float[num_inputs*num_outputs];
if ((weights==0)||(output_errors==0)||(back_errors==0)
    ||(outputs==0)||(expected_values==0))
```

```
            ||(past_deltas==0)||(cum_deltas==0))
            {
            cout << "not enough memory\n";
            cout << "choose a smaller architecture\n";
            exit(1);
            }
// zero cum_deltas and past_deltas matrix
for (i=0; i< num_inputs; i++)
        {
        k=i*num_outputs;
        for (j=0; j< num_outputs; j++)
            {
            cum_deltas[k+j]=0;
            past_deltas[k+j]=0;
            }
        }

}
```

The destructor simply deletes the new vectors:

```
output_layer::~output_layer()
{
// some compilers may require the array
// size in the delete statement; those
// conforming to Ansi C++ will not
delete [num_outputs*num_inputs] weights;
delete [num_outputs] output_errors;
delete [num_inputs] back_errors;
delete [num_outputs] outputs;
delete [num_outputs*num_inputs] past_deltas;
delete [num_outputs*num_inputs] cum_deltas;
}
```

Now let's look at the **update_weights()** routine changes:

```
void output_layer::update_weights(const float beta,
                const float alpha)
{
int i, j, k;
float delta;
// learning law: weight_change =
//     beta*output_error*input + alpha*past_delta
for (i=0; i< num_inputs; i++)
    {
    k=i*num_outputs;
    for (j=0; j< num_outputs; j++)
        {
        delta=beta*output_errors[j]*(*(inputs+i))
            +alpha*past_deltas[k+j];
        weights[k+j] += delta;
```

```
      cum_deltas[k+j]+=delta; // current cycle
      }

   }

}
```

The change to the training law amounts to calculating a delta and adding it to the cumulative total of weight changes in **cum_deltas**. At some point (at the start of a new cycle) you need to set the **past_deltas** vector to the **cum_delta** vector. Where does this occur? Since the layer has no concept of cycle, this must be done at the network level. There is a network level function called **update_momentum** at the beginning of each cycle that in turns calls a layer level function of the same name. The layer level function swaps the **past_deltas** vector and the **cum_deltas** vector, and reinitializes the **cum_deltas** vector to zero. We need to return to the layer.h file to see changes that are needed to define the two functions mentioned.

```
class output_layer: public layer
{
protected:

    float * weights;
    float * output_errors; // array of errors at output
    float * back_errors; // array of errors back-propagated
    float * expected_values;   // to inputs
    float * cum_deltas; // for momentum
    float * past_deltas;// for momentum

   friend network;

public:

    output_layer(int, int);
    ~output_layer();
    virtual void calc_out();
    void calc_error(float &);
    void randomize_weights();
    void update_weights(const float, const float);
    void update_momentum();
    void list_weights();
    void write_weights(int, FILE *);
    void read_weights(int, FILE *);
    void list_errors();
    void list_outputs();
};
```

```
class network

{

private:

layer *layer_ptr[MAX_LAYERS];
    int number_of_layers;
    int layer_size[MAX_LAYERS];
    float *buffer;
    fpos_t position;
    unsigned training;

public:
  network();
    ~network();
      void set_training(const unsigned &);
      unsigned get_training_value();
      void get_layer_info();
      void set_up_network();
      void randomize_weights();
      void update_weights(const float, const float);
      void update_momentum();
      ...
```

At both the **Network** and **output_layer** class levels the function prototype for the **update_momentum** member functions are highlighted. The implementation for these functions are shown as follows from the **layer.cpp** class.

```
void output_layer::update_momentum()
{
// This function is called when a
// new cycle begins; the past_deltas
// pointer is swapped with the
// cum_deltas pointer. Then the contents
// pointed to by the cum_deltas pointer
// is zeroed out.
int i, j, k;
float * temp;

// swap
temp - past_deltas;
past_deltas=cum_deltas;
cum_deltas=temp;

// zero cum_deltas matrix
```

```
// for new cycle
for (i=0; i< num_inputs; i++)
    {
    k=i*num_outputs;
    for (j=0; j< num_outputs; j++)
        cum_deltas[k+j]=0;
    }
}

void network::update_momentum()
{
int i;

for (i=1; i<number_of_layers; i++)
    ((output_layer *)layer_ptr[i])
        ->update_momentum();
}
```

Adding Noise During Training

Another approach to breaking out of local minima is to introduce some noise in the inputs during training. A random number is added to each input component of the input vector as it is applied to the network. This is scaled by an overall noise factor, NF, which has a 0 to 1 range. You can add as much noise to the simulation as you want, or not any at all, by choosing NF = 0. When you are close to a solution and have reached a satisfactory minimum, you don't want noise at that time to interfere with convergence to the minimum. We implement a noise factor that decreases with the number of cycles as shown in the following excerpt from the backprop.cpp file.

```
// update NF
// gradually reduce noise to zero
if (total_cycles>0.7*max_cycles)
                    new_NF = 0;
else if (total_cycles>0.5*max_cycles)
                    new_NF = 0.25*NF;
else if (total_cycles>0.3*max_cycles)
                    new_NF = 0.50*NF;
else if (total_cycles>0.1*max_cycles)
                    new_NF = 0.75*NF;

backp.set_NF(new_NF);
```

The noise factor is reduced at regular intervals. The new noise factor is updated with the network class function called **set_NF(float)**. There is a member variable in the network class called **NF** that holds the current value for the noise factor. The noise is added to the inputs in the **input_layer** member function **calc_out()**.

Another reason for using noise is to prevent memorization by the network. You are effectively presenting a different input pattern with each cycle so it becomes hard for the network to memorize patterns.

One Other Change—Starting Training from a Saved Weight File

Shortly, we will look at the complete listings for the backpropagation simulator. There is one other enhancement to discuss. It is often useful in long simulations to be able to start from a known point, which is from an already saved set of weights. This is a simple change in the backprop.cpp program, which is well worth the effort. As a side benefit, this feature will allow you to run a simulation with a large beta value for, say, 500 cycles, save the weights, and then start a new simulation with a smaller beta value for another 500 or more cycles. You can take preset breaks in long simulations, which you will encounter in Chapter 14. At this point, let's look at the complete listings for the updated layer.h and layer.cpp files in Listings 13.1 and 13.2:

Listing 13.1   layer.h file updated to include noise and momentum

```
// layer.h    V.Rao, H. Rao
// header file for the layer class hierarchy and
// the network class
 // added noise and momentum

#define MAX_LAYERS   5
#define MAX_VECTORS 100


class network;
class Kohonen_network;

class layer
{

protected:

    int num_inputs;
    int num_outputs;
```

```
    float *outputs; // pointer to array of outputs
    float *inputs;  // pointer to array of inputs, which
                    // are outputs of some other layer

    friend network;
    friend Kohonen_network; // update for Kohonen model

public:

    virtual void calc_out()=0;
};


class input_layer: public layer
{

private:

float noise_factor;
float * orig_outputs;

public:

    input_layer(int, int);
    ~input_layer();
    virtual void calc_out();
    void set_NF(float);

    friend network;
};

class middle_layer;

class output_layer: public layer
{
protected:

    float * weights;
    float * output_errors; // array of errors at output
    float * back_errors; // array of errors back-propagated
    float * expected_values;   // to inputs
    float * cum_deltas; // for momentum
    float * past_deltas;// for momentum

     friend network;

public:

    output_layer(int, int);
```

```
    ~output_layer();
    virtual void calc_out();
    void calc_error(float &);
    void randomize_weights();
    void update_weights(const float, const float);
    void update_momentum();
    void list_weights();
    void write_weights(int, FILE *);
    void read_weights(int, FILE *);
    void list_errors();
    void list_outputs();
};


class middle_layer: public output_layer
{

private:

public:
    middle_layer(int, int);
    ~middle_layer();
    void calc_error();
};


class network

{

private:

layer *layer_ptr[MAX_LAYERS];
    int number_of_layers;
    int layer_size[MAX_LAYERS];
    float *buffer;
    fpos_t position;
    unsigned training;

public:
    network();
    ~network();
        void set_training(const unsigned &);
        unsigned get_training_value();
        void get_layer_info();
        void set_up_network();
        void randomize_weights();
        void update_weights(const float, const float);
        void update_momentum();
```

```
            void write_weights(FILE *);
            void read_weights(FILE *);
            void list_weights();
            void write_outputs(FILE *);
            void list_outputs();
            void list_errors();
            void forward_prop();
            void backward_prop(float &);
            int fill_IObuffer(FILE *);
            void set_up_pattern(int);
            void set_NF(float);

};
```

Listing 13.2   Layer.cpp file updated to include noise and momentum

```
// layer.cpp      V.Rao, H.Rao
// added momentum and noise

// compile for floating point hardware if available
#include <stdio.h>
#include <iostream.h>
#include <stdlib.h>
#include <math.h>
#include <time.h>
#include "layer.h"


inline float squash(float input)
// squashing function
// use sigmoid -- can customize to something
// else if desired; can add a bias term too
//
{
if (input < -50)
    return 0.0;
else   if (input > 50)
        return 1.0;
    else return (float)(1/(1+exp(-(double)input)));

}


inline float randomweight(unsigned init)
{
int num;
// random number generator
// will return a floating point
// value between -1 and 1

if (init==1) // seed the generator
    srand ((unsigned)time(NULL));
```

```
num=rand() % 100;

return 2*(float(num/100.00))-1;
}

// the next function is needed for Turbo C++
// and Borland C++ to link in the appropriate
// functions for fscanf floating point formats:
static void force_fpf()
{
    float x, *y;
    y=&x;
    x=*y;
}

// -----------------------------------------
//              input layer
//------------------------------------------
input_layer::input_layer(int i, int o)
{


num_inputs=i;
num_outputs=o;

outputs = new float[num_outputs];
orig_outputs = new float[num_outputs];
if ((outputs==0)||(orig_outputs==0))
    {
    cout << "not enough memory\n";
    cout << "choose a smaller architecture\n";
    exit(1);
    }

noise_factor=0;

}

input_layer::~input_layer()
{
delete [num_outputs] outputs;
delete [num_outputs] orig_outputs;
}


void input_layer::calc_out()
{
//add noise to inputs
// randomweight returns a random number
```

```
// between -1 and 1

int i;
for (i=0; i<num_outputs; i++)
    outputs[i] =orig_outputs[i]*
        (1+noise_factor*randomweight(0));

}

void input_layer::set_NF(float noise_fact)
{
noise_factor=noise_fact;
}



// ------------------------------------------
//              output layer
//-------------------------------------------



output_layer::output_layer(int ins, int outs)
{
int i, j, k;
num_inputs=ins;
num_outputs=outs;
weights = new float[num_inputs*num_outputs];
output_errors = new float[num_outputs];
back_errors = new float[num_inputs];
outputs = new float[num_outputs];
expected_values = new float[num_outputs];
cum_deltas = new float[num_inputs*num_outputs];
past_deltas = new float[num_inputs*num_outputs];

if ((weights==0)||(output_errors==0)||(back_errors==0)
    ||(outputs==0)||(expected_values==0)
    ||(past_deltas==0)||(cum_deltas==0))
    {
    cout << "not enough memory\n";
    cout << "choose a smaller architecture\n";
    exit(1);
    }

// zero cum_deltas and past_deltas matrix
for (i=0; i< num_inputs; i++)
    {
    k=i*num_outputs;
    for (j=0; j< num_outputs; j++)
```

```
          {
          cum_deltas[k+j]=0;
          past_deltas[k+j]=0;
          }
      }
}

output_layer::~output_layer()
{
// some compilers may require the array
// size in the delete statement; those
// conforming to Ansi C++ will not
delete [num_outputs*num_inputs] weights;
delete [num_outputs] output_errors;
delete [num_inputs] back_errors;
delete [num_outputs] outputs;
delete [num_outputs*num_inputs] past_deltas;
delete [num_outputs*num_inputs] cum_deltas;

}


void output_layer::calc_out()
{

int i,j,k;
float accumulator=0.0;

for (j=0; j<num_outputs; j++)
    {

    for (i=0; i<num_inputs; i++)

        {
        k=i*num_outputs;
        if (weights[k+j]*weights[k+j] > 1000000.0)
            {
            cout << "weights are blowing up\n";
            cout << "try a smaller learning constant\n";
            cout << "e.g. beta=0.02    aborting...\n";
            exit(1);
            }
        outputs[j]=weights[k+j]*(*(inputs+i));
        accumulator+=outputs[j];
        }
    // use the sigmoid squash function
    outputs[j]=squash(accumulator);
    accumulator=0;
    }
```

```
}


void output_layer::calc_error(float & error)
{
int i, j, k;
float accumulator=0;
float total_error=0;

for (j=0; j<num_outputs; j++)
    {
        output_errors[j] = expected_values[j]-outputs[j];
        total_error+=output_errors[j];
        }

error=total_error;

for (i=0; i<num_inputs; i++)
{
k=i*num_outputs;
for (j=0; j<num_outputs; j++)
    {
        back_errors[i]=
            weights[k+j]*output_errors[j];
        accumulator+=back_errors[i];
        }
    back_errors[i]=accumulator;
    accumulator=0;
    // now multiply by derivative of
    // sigmoid squashing function, which is
    // just the input*(1-input)
    back_errors[i]*=(*(inputs+i))*(1-(*(inputs+i)));
    }

}
void output_layer::randomize_weights()
{
int i, j, k;
const unsigned first_time=1;

const unsigned not_first_time=0;
float discard;

discard=randomweight(first_time);

for (i=0; i< num_inputs; i++)
    {
    k=i*num_outputs;
    for (j=0; j< num_outputs; j++)
```

```
            weights[k+j]=randomweight(not_first_time);
      }
}

void output_layer::update_weights(const float beta,
                  const float alpha)
{
int i, j, k;
float delta;

// learning law: weight_change =
//      beta*output_error*input + alpha*past_delta

for (i=0; i< num_inputs; i++)
    {
    k=i*num_outputs;
    for (j=0; j< num_outputs; j++)
        {
        delta=beta*output_errors[j]*(*(inputs+i))
            +alpha*past_deltas[k+j];
        weights[k+j] += delta;
        cum_deltas[k+j]+=delta; // current cycle
        }

    }

}

void output_layer::update_momentum()
{
// This function is called when a
// new cycle begins; the past_deltas
// pointer is swapped with the
// cum_deltas pointer. Then the contents
// pointed to by the cum_deltas pointer
// is zeroed out.
int i, j, k;
float * temp;

// swap
temp = past_deltas;
past_deltas=cum_deltas;
cum_deltas=temp;

// zero cum_deltas matrix
// for new cycle
for (i=0; i< num_inputs; i++)
    {
    k=i*num_outputs;
    for (j=0; j< num_outputs; j++)
        cum_deltas[k+j]=0;
```

```
    }
}

void output_layer::list_weights()
{
int i, j, k;

for (i=0; i< num_inputs; i++)
    {
    k=i*num_outputs;
    for (j=0; j< num_outputs; j++)
        cout << "weight["<<i<<","<<
            j<<"] is: "<<weights[k+j];
    }

}

void output_layer::list_errors()
{
int i, j;

for (i=0; i< num_inputs; i++)
    cout << "backerror["<<i<<
        "] is : "<<back_errors[i]<<"\n";

for (j=0; j< num_outputs; j++)
    cout << "outputerrors["<<j<<
            "] is: "<<output_errors[j]<<"\n";

}


void output_layer::write_weights(int layer_no,
        FILE * weights_file_ptr)
{
int i, j, k;

// assume file is already open and ready for
// writing

// prepend the layer_no to all lines of data
// format:
//      layer_no  weight[0,0] weight[0,1] ...
//      layer_no  weight[1,0] weight[1,1] ...
//      ...

for (i=0; i< num_inputs; i++)
    {
    fprintf(weights_file_ptr,"%i ",layer_no);
    k=i*num_outputs;
     for (j=0; j< num_outputs; j++)
```

```
        {
        fprintf(weights_file_ptr,"%f ",
                weights[k+j]);
        }
      fprintf(weights_file_ptr,"\n");
      }


}

void output_layer::read_weights(int layer_no,
      FILE * weights_file_ptr)
{
int i, j, k;


// assume file is already open and ready for
// reading

// look for the prepended layer_no
// format:
//    layer_no  weight[0,0] weight[0,1] ...
//    layer_no  weight[1,0] weight[1,1] ...
//    ...
while (1)

    {

    fscanf(weights_file_ptr,"%i",&j);
    if ((j==layer_no)|| (feof(weights_file_ptr)))
       break;
    else
       {
       while (fgetc(weights_file_ptr) != '\n')
          {;}// get rest of line
       }
    }

if (!(feof(weights_file_ptr)))
   {
   // continue getting first line
   i=0;
   for (j=0; j< num_outputs; j++)
         {

         fscanf(weights_file_ptr,"%f",
               &weights[j]); // i*num_outputs = 0
      }
   fscanf(weights_file_ptr,"\n");
```

```
    // now get the other lines
    for (i=1; i< num_inputs; i++)
        {
        fscanf(weights_file_ptr,"%i",&layer_no);
    k=i*num_outputs;
    for (j=0; j< num_outputs; j++)
        {
        fscanf(weights_file_ptr,"%f",
            &weights[k+j]);
            }

    }
    fscanf(weights_file_ptr,"\n");
    }


else cout << "end of file reached\n";

}

void output_layer::list_outputs()
{
int j;

for (j=0; j< num_outputs; j++)
    {
    cout << "outputs["<<j
        <<"] is: "<<outputs[j]<<"\n";
    }

}


// ----------------------------------------
//              middle layer
//-----------------------------------------


middle_layer::middle_layer(int i, int o):
    output_layer(i,o)
{

}

middle_layer::~middle_layer()
{
delete [num_outputs*num_inputs] weights;
delete [num_outputs] output_errors;
delete [num_inputs] back_errors;
```

```
delete [num_outputs] outputs;
}


void middle_layer::calc_error()
{
int i, j, k;
float accumulator=0;

for (i=0; i<num_inputs; i++)
    {
    k=i*num_outputs;
    for (j=0; j<num_outputs; j++)
        {
        back_errors[i]=
            weights[k+j]*(*(output_errors+j));
        accumulator+=back_errors[i];
        }
    back_errors[i]=accumulator;
    accumulator=0;
    // now multiply by derivative of
    // sigmoid squashing function, which is
    // just the input*(1-input)
    back_errors[i]*=(*(inputs+i))*(1-(*(inputs+i)));
    }

}

network::network()
{
position=0L;
}

network::~network()
{
int i,j,k;
i=layer_ptr[0]->num_outputs;// inputs
j=layer_ptr[number_of_layers-1]->num_outputs; //outputs
k=MAX_VECTORS;


delete [(i+j)*k]buffer;
}

void network::set_training(const unsigned & value)
{
training=value;
}

unsigned network::get_training_value()
{
```

```
return training;
}


void network::get_layer_info()
{
int i;

//--------------------------------------------
//
// Get layer sizes for the network
//
// --------------------------------------------



cout << " Please enter in the number of layers for your net\
    work.\n";
cout << " You can have a minimum of 3 to a maximum of 5. \n";
cout << " 3 implies 1 hidden layer; 5 implies 3 hidden \
    layers : \n\n";

cin >> number_of_layers;

cout << " Enter in the layer sizes separated by spaces.\n";
cout << " For a network with 3 neurons in the input \
    layer,\n";
cout << " 2 neurons in a hidden layer, and 4 neurons in \
    the\n";
cout << " output layer, you would enter: 3 2 4 .\n";
cout << " You can have up to 3 hidden layers,for five \
    maximum entries :\n\n";

for (i=0; i<number_of_layers; i++)
    {
    cin >> layer_size[i];
    }

// ----------------------------------------------------------
// size of layers:
//     input_layer        layer_size[0]
//     output_layer       layer_size[number_of_layers-1]
//     middle_layers      layer_size[1]
//     optional: layer_size[number_of_layers-3]
//     optional: layer_size[number_of_layers-2]
//----------------------------------------------------------


}

void network::set_up_network()
```

```
{
int i,j,k;
//----------------------------------------------------------
// Construct the layers
//
//----------------------------------------------------------


layer_ptr[0] = new input_layer(0,layer_size[0]);

for (i=0;i<(number_of_layers-1);i++)
    {
    layer_ptr[i+1] =
    new middle_layer(layer_size[i],layer_size[i+1]);
    }

layer_ptr[number_of_layers-1] = new
output_layer(layer_size[number_of_layers-2],
    layer_size[number_of_layers-1]);

for (i=0;i<(number_of_layers-1);i++)
    {
    if (layer_ptr[i] == 0)
        {
        cout << "insufficient memory\n";
        cout << "use a smaller architecture\n";
        exit(1);
        }
    }

//----------------------------------------------------------
// Connect the layers
//
//----------------------------------------------------------
// set inputs to previous layer outputs for all layers,
//     except the input layer

for (i=1; i< number_of_layers; i++)
    layer_ptr[i]->inputs = layer_ptr[i-1]->outputs;


// for back_propagation, set output_errors to next layer
//     back_errors for all layers except the output
//     layer and input layer


for (i=1; i< number_of_layers -1; i++)
    ((output_layer *)layer_ptr[i])->output_errors =
        ((output_layer *)layer_ptr[i+1])->back_errors;
```

```
// define the IObuffer that caches data from
// the datafile
i=layer_ptr[0]->num_outputs;// inputs
j=layer_ptr[number_of_layers-1]->num_outputs; //outputs
k=MAX_VECTORS;

buffer=new
    float[(i+j)*k];
if (buffer==0)
    {
    cout << "insufficient memory for buffer\n";
    exit(1);
    }
}

void network::randomize_weights()
{
int i;

for (i=1; i<number_of_layers; i++)
    ((output_layer *)layer_ptr[i])
        ->randomize_weights();
}


void network::update_weights(const float beta, const float
    alpha)
{
int i;

for (i=1; i<number_of_layers; i++)
    ((output_layer *)layer_ptr[i])
        ->update_weights(beta,alpha);
}

void network::update_momentum()
{
int i;

for (i=1; i<number_of_layers; i++)
    ((output_layer *)layer_ptr[i])
        ->update_momentum();
}


void network::write_weights(FILE * weights_file_ptr)
{
int i;

for (i=1; i<number_of_layers; i++)
```

```
    ((output_layer *)layer_ptr[i])
        ->write_weights(i,weights_file_ptr);
}


void network::read_weights(FILE * weights_file_ptr)
{
int i;

for (i=1; i<number_of_layers; i++)
    ((output_layer *)layer_ptr[i])
        ->read_weights(i,weights_file_ptr);
}


void network::list_weights()
{
int i;

for (i=1; i<number_of_layers; i++)
    {
    cout << "layer number : " <<i<< "\n";
    ((output_layer *)layer_ptr[i])
        ->list_weights();
    }
}

void network::list_outputs()
{
int i;

for (i=1; i<number_of_layers; i++)
    {
    cout << "layer number : " <<i<< "\n";
    ((output_layer *)layer_ptr[i])
        ->list_outputs();
    }
}

void network::write_outputs(FILE *outfile)
{
int i, ins, outs;
ins=layer_ptr[0]->num_outputs;
outs=layer_ptr[number_of_layers-1]->num_outputs;
float temp;

fprintf(outfile,"for input vector:\n");

for (i=0; i<ins; i++)
    {
    temp=layer_ptr[0]->outputs[i];
```

```
        fprintf(outfile,"%f  ",temp);
        }


    fprintf(outfile,"\noutput vector is:\n");

    for (i=0; i<outs; i++)
        {
        temp=layer_ptr[number_of_layers-1]->
        outputs[i];
        fprintf(outfile,"%f  ",temp);

        }

    if (training==1)
    {
    fprintf(outfile,"\nexpected output vector is:\n");

    for (i=0; i<outs; i++)
        {
        temp=((output_layer *)(layer_ptr[number_of_layers-1]))->
        expected_values[i];
        fprintf(outfile,"%f  ",temp);

        }
    }

    fprintf(outfile,"\n---------------------\n");

    }



    void network::list_errors()
    {
    int i;

    for (i=1; i<number_of_layers; i++)
        {
        cout << "layer number : " <<i<< "\n";
        ((output_layer *)layer_ptr[i])
            ->list_errors();
        }
    }



    int network::fill_IObuffer(FILE * inputfile)
    {
    // this routine fills memory with
    // an array of input, output vectors
```

```
// up to a maximum capacity of
// MAX_INPUT_VECTORS_IN_ARRAY
// the return value is the number of read
// vectors


int i, k, count, veclength;

int ins, outs;

ins=layer_ptr[0]->num_outputs;

outs=layer_ptr[number_of_layers-1]->num_outputs;

if (training==1)
    veclength=ins+outs;
else
    veclength=ins;

count=0;
while  ((count<MAX_VECTORS)&&
        (!feof(inputfile)))
    {
    k=count*(veclength);
    for (i=0; i<veclength; i++)
        {
        fscanf(inputfile,"%f",&buffer[k+i]);
        }
    fscanf(inputfile,"\n");
    count++;
    }

if (!(ferror(inputfile)))
    return count;
else return -1; // error condition

}


void network::set_up_pattern(int buffer_index)
{
// read one vector into the network
int i, k;
int ins, outs;

ins=layer_ptr[0]->num_outputs;
outs=layer_ptr[number_of_layers-1]->num_outputs;
if (training==1)
    k=buffer_index*(ins+outs);
else
    k=buffer_index*ins;
```

```
for (i=0; i<ins; i++)
    ((input_layer*)layer_ptr[0])
            ->orig_outputs[i]=buffer[k+i];

if (training==1)
{
    for (i=0; i<outs; i++)

        ((output_layer *)layer_ptr[number_of_layers-1])->
            expected_values[i]=buffer[k+i+ins];
}

}


void network::forward_prop()
{
int i;
for (i=0; i<number_of_layers; i++)
    {
    layer_ptr[i]->calc_out(); //polymorphic
            // function
    }
}

void network::backward_prop(float & toterror)
{
int i;

// error for the output layer
((output_layer*)layer_ptr[number_of_layers-1])->
            calc_error(toterror);

// error for the middle layer(s)
for (i=number_of_layers-2; i>0; i--)
    {
    ((middle_layer*)layer_ptr[i])->
            calc_error();

    }

}

void network::set_NF(float noise_fact)
{
((input_layer*)layer_ptr[0])->set_NF(noise_fact);
}
```

The New and Final backprop.cpp File

The last file to present is the backprop.cpp file. This is shown in Listing 13.3:

Listing 13.3   Implementation file for the backpropagation simulator, with noise and momentum backprop.cpp

```
// backprop.cpp      V. Rao, H. Rao
#include "layer.cpp"

#define TRAINING_FILE   "training.dat"
#define WEIGHTS_FILE "weights.dat"
#define OUTPUT_FILE "output.dat"
#define TEST_FILE   "test.dat"

void main()
{

float error_tolerance=0.1;
float total_error=0.0;
float avg_error_per_cycle=0.0;
float error_last_cycle=0.0;
float avgerr_per_pattern=0.0; // for the latest cycle
float error_last_pattern=0.0;
float learning_parameter=0.02;
float alpha; // momentum parameter
float NF; // noise factor
float new_NF;

unsigned temp, startup, start_weights;
long int vectors_in_buffer;
long int max_cycles;
long int patterns_per_cycle=0;

long int total_cycles, total_patterns;
int i;


// create a network object
network backp;


FILE * training_file_ptr, * weights_file_ptr, *
    output_file_ptr;
FILE * test_file_ptr, * data_file_ptr;

// open output file for writing
if ((output_file_ptr=fopen(OUTPUT_FILE,"w"))==NULL)
        {
        cout << "problem opening output file\n";
```

```
        exit(1);
        }

// enter the training mode : 1=training on      0=training off
cout << "----------------------------------------------------
    \n";
cout << " C++ Neural Networks and Fuzzy Logic \n";
cout << " Backpropagation simulator \n";
cout << "      version 2 \n";
cout << "----------------------------------------------------
    \n";
cout << "Please enter 1 for TRAINING on, or 0 for off: \n\n";
cout << "Use training to change weights according to your\n";
cout << "expected outputs. Your training.dat file should con\
    tain\n";
cout << "a set of inputs and expected outputs. The number \
    of\n";
cout << "inputs determines the size of the first (input) \
    layer\n";
cout << "while the number of outputs determines the size of \
    the\n";
cout << "last (output) layer :\n\n";

cin >> temp;
backp.set_training(temp);

if (backp.get_training_value() == 1)
    {
    cout << "--> Training mode is *ON*. weights will be \
    saved\n";
    cout << "in the file weights.dat at the end of the\n";
    cout << "current set of input (training) data\n";
    }
else
    {
    cout << "--> Training mode is *OFF*. weights will be \
    loaded\n";
    cout << "from the file weights.dat and the current\n";
    cout << "(test) data set will be used. For the test\n";
    cout << "data set, the test.dat file should contain\n";
    cout << "only inputs, and no expected outputs.\n";
    }

if (backp.get_training_value()==1)
    {
    // ------------------------------------------
    // Read in values for the error_tolerance,
    // and the learning_parameter
    // ------------------------------------------
    cout << " Please enter in the error_tolerance\n";
    cout << " --- between 0.001 to 100.0, try 0.1 to start -\
```

```
\n";
cout << "\n";
cout << "and the learning_parameter, beta\n";
cout << " --- between 0.01 to 1.0, try 0.5 to start -\
\n\n";
cout << " separate entries by a space\n";
cout << " example: 0.1 0.5 sets defaults mentioned :\n\n";

cin >> error_tolerance >> learning_parameter;

// -------------------------------------------
// Read in values for the momentum
// parameter, alpha (0-1.0)
// and the noise factor, NF (0-1.0)
// -------------------------------------------
cout << "Enter values now for the momentum \n";
cout << "parameter, alpha(0-1.0)\n";
cout << " and the noise factor, NF (0-1.0)\n";
cout << "You may enter zero for either of these\n";
cout << "parameters, to turn off the momentum or\n";
cout << "noise features.\n";
cout << "If the noise feature is used, a random\n";
cout << "component of noise is added to the inputs\n";
cout << "This is decreased to 0 over the maximum\n";
cout << "number of cycles specified.\n";
cout << "enter alpha followed by NF, e.g., 0.3 0.5\n";

cin >> alpha >> NF;

//-------------------------------------------
// open training file for reading
//-------------------------------------------
if ((training_file_ptr=fopen(TRAINING_FILE,"r"))==NULL)
    {
    cout << "problem opening training file\n";
    exit(1);
    }
data_file_ptr=training_file_ptr; // training on

// Read in the maximum number of cycles
// each pass through the input data file is a cycle
cout << "Please enter the maximum cycles for the simula\
tion\n";
cout << "A cycle is one pass through the data set.\n";
cout << "Try a value of 10 to start with\n";

cin >> max_cycles;

cout << "Do you want to read weights from weights.dat to \
start?\n";
cout << "Type 1 to read from file, 0 to randomize \
```

```
        starting weights\n";
        cin >> start_weights;

        }
    else
        {
        if ((test_file_ptr=fopen(TEST_FILE,"r"))==NULL)
            {
            cout << "problem opening test file\n";
            exit(1);
            }

        data_file_ptr=test_file_ptr; // training off
        }



// training: continue looping until the total error is less
// than
//      the tolerance specified, or the maximum number of
//      cycles is exceeded; use both the forward signal propa
//      gation
//      and the backward error propagation phases. If the error
//      tolerance criteria is satisfied, save the weights in a
//      file.
// no training: just proceed through the input data set once
// in the
//      forward signal propagation phase only. Read the start
//      ing
//      weights from a file.
// in both cases report the outputs on the screen


// initialize counters
total_cycles=0; // a cycle is once through all the input data
total_patterns=0; // a pattern is one entry in the input data
new_NF=NF;


// get layer information
backp.get_layer_info();

// set up the network connections
backp.set_up_network();

// initialize the weights
if ((backp.get_training_value()==1)&&(start_weights!=1))
    {
    // randomize weights for all layers; there is no
    // weight matrix associated with the input layer
```

```
        // weight file will be written after processing

        backp.randomize_weights();
        // set up the noise factor value
        backp.set_NF(new_NF);
        }
    else
        {
        // read in the weight matrix defined by a
        // prior run of the backpropagation simulator
        // with training on
        if ((weights_file_ptr=fopen(WEIGHTS_FILE,"r"))
                ==NULL)
            {
            cout << "problem opening weights file\n";
            exit(1);
            }
        backp.read_weights(weights_file_ptr);
        fclose(weights_file_ptr);
        }




    // main loop
    // if training is on, keep going through the input data
    //      until the error is acceptable or the maximum number of
    //      cycles
    //      is exceeded.
    // if training is off, go through the input data once. report
    // outputs
    // with inputs to file output.dat

    startup=1;
    vectors_in_buffer = MAX_VECTORS; // startup condition
    total_error = 0;


    while (    ((backp.get_training_value()==1)
            && (avgerr_per_pattern
                    > error_tolerance)
            && (total_cycles < max_cycles)
            && (vectors_in_buffer !=0))
            || ((backp.get_training_value()==0)
            && (total_cycles < 1))
            || ((backp.get_training_value()==1)
            && (startup==1))
            )
    {
    startup=0;
    error_last_cycle=0; // reset for each cycle
```

```
patterns_per_cycle=0;

backp.update_momentum(); // added to reset
            // momentum matrices
            // each cycle

// process all the vectors in the datafile
// going through one buffer at a time
// pattern by pattern


while ((vectors_in_buffer==MAX_VECTORS))
    {

    vectors_in_buffer=
        backp.fill_IObuffer(data_file_ptr); // fill buffer
        if (vectors_in_buffer < 0)
            {
            cout << "error in reading in vectors, aborting\n";
            cout << "check that there are on extra \
            linefeeds\n";
            cout << "in your data file, and that the number\n";
            cout << "of layers and size of layers match the\n";
            cout << "the parameters provided.\n";
            exit(1);
            }

        // process vectors
        for (i=0; i<vectors_in_buffer; i++)
            {
            // get next pattern
            backp.set_up_pattern(i);

            total_patterns++;
            patterns_per_cycle++;
            // forward propagate

            backp.forward_prop();

            if (backp.get_training_value()==0)
                backp.write_outputs(output_file_ptr);

            // back_propagate, if appropriate
            if (backp.get_training_value()==1)
                {

                backp.backward_prop(error_last_pattern);
                error_last_cycle +=
                    error_last_pattern*error_last_pattern;
                avgerr_per_pattern=
```

```
        ((float)sqrt((double)error_last_cycle))/patterns_per_cycle
        ;

                // if it's not the last cycle, update weights
                if ((avgerr_per_pattern
                    > error_tolerance)
                    && (total_cycles+1 < max_cycles))

                    backp.update_weights(learning_parameter,
                        alpha);
                // backp.list_weights(); // can
                // see change in weights by
                // using list_weights before and
                // after back_propagation
                }

            }

    error_last_pattern = 0;
        }

total_error += error_last_cycle;
total_cycles++;

// update NF
// gradually reduce noise to zero
if (total_cycles>0.7*max_cycles)
        new_NF = 0;
else    if (total_cycles>0.5*max_cycles)
            new_NF = 0.25*NF;
        else    if (total_cycles>0.3*max_cycles)
                new_NF = 0.50*NF;
            else    if (total_cycles>0.1*max_cycles)
                    new_NF = 0.75*NF;

backp.set_NF(new_NF);

// most character displays are 25 lines
// user will see a corner display of the cycle count
// as it changes

cout << "\n\n\n\n\n\n\n\n\n\n\n\n\n\n\n\n\n\n\n\n\n\n\n\n";
cout << total_cycles << "\t" << avgerr_per_pattern << "\n";

fseek(data_file_ptr, 0L, SEEK_SET); // reset the file pointer
            // to the beginning of
            // the file
vectors_in_buffer = MAX_VECTORS; // reset

} // end main loop
```

```
if (backp.get_training_value()==1)
    {
    if ((weights_file_ptr=fopen(WEIGHTS_FILE,"w"))
        ==NULL)
        {
        cout << "problem opening weights file\n";
        exit(1);
        }
    }

cout << "\n\n\n\n\n\n\n\n\n\n\n";
cout << "----------------------------------------------------\n";
cout << " done:  results in file output.dat\n";
cout << "     training: last vector only\n";
cout << "     not training: full cycle\n\n";
if (backp.get_training_value()==1)
    {
    backp.write_weights(weights_file_ptr);
    backp.write_outputs(output_file_ptr);
    avg_error_per_cycle=(float)sqrt((double)total_error)/
    total_cycles;
    error_last_cycle=(float)sqrt((double)error_last_cycle);
    fclose(weights_file_ptr);

cout << "     weights saved in file weights.dat\n";
cout << "\n";
cout << "---->average error per cycle = " <<
    avg_error_per_cycle << " <---\n";
cout << "---->error last cycle = " << error_last_cycle << "
    <---\n";
cout << "->error last cycle per pattern= " <<
    avgerr_per_pattern << " <---\n";

    }

cout << "----------->total cycles = " << total_cycles <<
    " <---\n";
cout << "----------->total patterns = " << total_patterns <<
    " <---\n";
cout << "----------------------------------------------------\n";
// close all files
fclose(data_file_ptr);
fclose(output_file_ptr);

    }
```

Trying the Noise and Momentum Features

You can test out the version 2 simulator, which you just compiled with the example that you saw at the beginning of the chapter. You will find that there is a lot of trial and error in finding optimum values for alpha, the noise factor, and beta. This is true also for the middle layer size and the number of middle layers. For some problems, the addition of momentum makes convergence much faster. For other problems, you may not find any noticeable difference. An example run of the five-character recognition problem discussed at the beginning of this chapter resulted in the following results with beta = 0.1, tolerance = 0.001, alpha = 0.25, NF = 0.1, and the layer sizes kept at 35 5 3.

```
- - - - - - - - - - - - - - - - - - - - - - - - - - - - - - - - - - - - - -
    done:  results in file output.dat
        training: last vector only
        not training: full cycle

    weights saved in file weights.dat

----->average error per cycle = 0.02993<---
----->error last cycle = 0.00498<---
->error last cycle per pattern= 0.000996 <---
----------->total cycles = 242 <---
----------->total patterns = 1210 <---
- - - - - - - - - - - - - - - - - - - - - - - - - - - - - - - - - - - - - -
```

NOTE

The network was able to converge on a better solution (in terms of error measurement) in one-fourth the number of cycles. You can try varying alpha and NF to see the effect on overall simulation time. You can now start from the same initial starting weights by specifying a value of 1 for the starting weights question. For large values of alpha and beta, the network usually will not converge, and the weights will get unacceptably large (you will receive a message to that effect).

Variations of the Backpropagation Algorithm

Backpropagation is a versatile neural network algorithm that very often leads to success. Its achilles heel is the slowness at which it converges for certain problems. Many variations of the algorithm exist in the literature to try to improve convergence speed and robustness. Variations have been proposed in the following portions of the algorithm:

◆ **Adaptive parameters.** You can set rules that modify alpha, the momentum parameter, and beta, the learning parameter, as the simulation progresses. For example, you can reduce beta whenever a weight change does not reduce the error. You can consider undoing the particular weight change, setting alpha to zero and redoing the weight change with the new value of beta.

◆ **Use other minimum search routines besides steepest descent.** For example, you could use Newton's method for finding a minimum, although this would be a fairly slow process.

◆ **Use different cost functions.** Instead of calculating the error (as expected—actual output), you could determine another cost function that you want to minimize.

◆ **Modify the architecture.** You could use partially connected layers instead of fully connected layers. Also, you can use a *recurrent* network, that is, one in which some outputs feed back as inputs.

Applications

Backpropagation remains the king of neural network architectures because of its ease of use and wide applicability. A few of the notable applications in the literature will be cited as examples.

◆ **NETTalk** In 1987, Sejnowski and Rosenberg developed a network that was connected to a speech synthesizer that was able to utter English words, being trained on phonemes. The architecture consisted of an input layer window of seven characters. The characters were part of English text that was scrolled by. The network was trained to pronounce the letter at the center of the window. The middle layer had 80 neurons, while the output layer consisted of 26 neurons. With 1,024 training patterns and 10 cycles, the network started making intelligible speech, similar to the process of a child learning to talk. After 50 cycles, the network was about 95% accurate. You could purposely damage the network with the removal of neurons, but this did not cause performance to drop off a cliff; instead, the performance degraded gracefully. There was rapid recovery with retraining with fewer neurons also. This shows the fault tolerance of neural networks.

◆ **Sonar target recognition.** Neural nets using backpropagation have been used to identify different types of targets using the frequency signature (with a Fast Fourier transform) of the reflected signal.

◆ **Car navigation.** Pomerleau developed a neural network that is able to navigate a car based on images obtained from a camera mounted on the car's roof, and a range finder that coded distances in gray scale. The 30 x 32 pixel image and the 8 x 32 range finder image were fed into a hidden layer of size 29 feeding an output layer of 45 neurons. The output neurons were arranged in a straight line with each side representing a turn to a particular direction (right or left), while the center neurons represented "drive straight ahead." After 1,200 road images were trained on the network, the neural network driver was able to negotiate a part of the Carnegie-Mellon campus at a speed of about 3 miles per hour, limited only by the speed of the real time calculations done on a trained network in the Sun-3 computer in the car.

◆ **Image compression.** G.W. Cottrell, P. Munro, and D. Zipser used backpropagation to compress images with the result of an 8:1 compression ratio. They used standard backpropagation with 64 input neurons (8 x 8 pixels), 16 hidden neurons, and 64 output neurons equal to the inputs. This is called *self-supervised backpropagation* and represents an autoassociative network. The compressed signal is taken from the hidden layer. The input to hidden layer comprised the compressor, while the hidden to output layer forms a decompressor.

◆ **Image recognition.** Le Cun reported a backpropagation network with three hidden layers that could recognize handwritten postal zip codes. He used a 16 x 16 array of pixel to represent each handwritten digit and needed to encode 10 outputs, each of which represented a digit from 0 to 9. One interesting aspect of this work is that the hidden layers were not fully connected. The network was set up with blocks of neurons in the first two hidden layers set up as feature detectors for different parts of the previous layer. All the neurons in the block were set up to have the same weights as those from the previous layer. This is called *weight sharing*. Each block would sample a different part of the previous layer's image. The first hidden layer had 12 blocks of 8 x 8 neurons, whereas the second hidden layer had 12 blocks of 4 x 4 neurons. The third hidden layer was fully connected and consisted of 30 neurons. There were 1,256 neurons. The network was trained on 7,300 examples and tested on 2,000 cases with error rates of 1% on training set and 5% on the test set.

Summary

- You explored further the backpropagation algorithm in this chapter, continuing the discussion in Chapter 7.

- A momentum term was added to the training law and was shown to result in much faster convergence in some cases.

- A noise term was added to inputs to allow training to take place with random noise applied. This noise was made to decrease with the number of cycles, so that final stage learning could be done in a noise-free environment.

- The final version of the backpropagation simulator was constructed and used on the example from Chapter 12. Further application of the simulator will be made in Chapter 14.

- Several applications with the backpropagation algorithm were outlined, showing the wide applicability of this algorithm.

Application to Financial Forecasting

n Chapters 7 and 13, the backpropagation simulator was developed. In this chapter, you will use the simulator to tackle a complex problem in financial forecasting. The application of neural networks to financial forecasting and modeling has been very popular over the last few years. Financial journals and magazines frequently mention the use of neural networks, and commercial tools and simulators are quite widespread.

Tuning a Network

There is no specific methodology for approaching problems with backpropagation. As you may have discovered, you have a number of choices to make for parameters and for the network itself:

1. The number of hidden layers.
2. The size of hidden layers.
3. The learning constant, beta(β).
4. The momentum parameter, alpha(α).
5. The range, format, and bias of data presented to the network.
6. The form of the squashing function (does not have to be the sigmoid).
7. The starting point, that is, initial weight matrix.
8. The addition of noise.

Some of the parameters listed can be made to vary with the number of cycles executed, similar to the current implementation of noise. For example, you can start with a learning constant β that is large and reduce this constant as learning progresses. This allows rapid initial learning in the beginning of the process and may speed up the overall simulation time.

Much of the process of determining the best parameters for a given application is trial and error. You need to spend a great deal of time evaluating different options to find the best fit for your problem. The importance of preconditioning your data for application to your network cannot be overemphasized. You will read more about this shortly.

Generalization

If your overall goal is beyond pattern classification, you need to keep track of your network's ability to generalize. Not only should you look at the overall error with the training set that you define, but you should retain some training examples as part of a test set, with which you can see whether or not the network is able to correctly master. If the network responds poorly to your test set, you know that you have overtrained, or you can say the network "memorized" the training patterns. If you look at the arbitrary curve-fitting analogy in Figure 14.1, you see curves for a generalized fit, labeled G, and an overfit, labeled O. In the case of the overfit, any data point outside of the training data results in highly erroneous prediction.

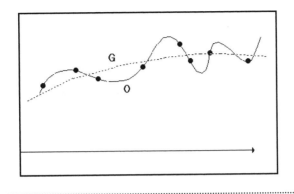

FIGURE 14.1 *General (G) versus over fitting (0) of data.*

Forecasting the S & P 500

The S & P 500 index is a widely followed stock average, like the Dow Jones Industrial Average (DJIA). It has a broader representation of the stock market since this average is based on 500 stocks, whereas the DJIA is based on only 30. The problem to be approached in this chapter is to predict the weekly change of the S & P 500 index, given a variety of indicators and data for prior weeks.

Choosing the Right Outputs

Whereas the objective may be to predict the level of the S & P 500, it is important to simplify the job of the network by asking for a weekly change in the level rather than for the absolute level of the index. What you want to do is to give the network the ability to fit the problem at hand conveniently in the output space of the output layer. Practically speaking, you know that the output from the network cannot be outside of the 0 to 1 range, since we have used a **Sigmoid Activation** function. You could take the S & P 500 index and scale this absolute price level to this range, for example. However you will likely end up with very small numbers that have a small range of variability. The difference from week to week, on the other hand, has a much smaller overall range, and when these differences are scaled to the 0 to 1 range, you have much more variability.

Choosing the Right Inputs

The inputs to the network need to be weekly changes of indicators that have some relevance to the S & P 500 index. This is a complex forecasting problem, and we can only guess at most of the relationships. This is one of the inherent strengths of using neural nets for forecasting; if a relationship is weak, the network will learn to ignore it automatically. In this example, we choose a data set that represents the state of the financial markets and the economy. The inputs chosen are listed as:

◆ Previous price action in the S & P 500 index, including the close or final value of the index for a given week and the high and low values achieved for the week.

◆ Breadth indicators for the stock market, including the number of advancing issues and declining issues for the stocks in the New York Stock Exchange (NYSE) and the much broader National Association of Securities Dealers (NASDAQ) market.

◆ Other technical indicators, including the number of new highs and new lows achieved in the week for the NYSE and NASDAQ markets. This gives some indication about the strength of an uptrend or downtrend.

◆ Trading volume statistics for the NYSE and NASDAQ markets, including the total volume for the week as well as the volume of all advancing issues, and the volume of all declining issues.

◆ Interest rates, including short-term interest rates in the Three-Month Treasury Bill Yield, and long-term rates in the 30-Year Treasury Bond Yield.

◆ The price of gold, which is usually correlated to the rate of inflation.

Government statistics like the Consumer Price Index, Housing starts, and the Unemployment Rate were not chosen specifically because of their monthly availability (versus this weekly forecast) and likelihood of having frequent revisions in the data. An important consideration to keep in mind when picking inputs is to try to avoid similar or highly correlated inputs. For example, choosing the price of crude oil would have been unnecessary since crude oil and gold are highly correlated to each other historically.

Choosing a Network Architecture

The input layer and output layer are fixed by the number of inputs and outputs we are using. In our case, the output is a single number, the expected change in the S & P 500 index in the next week. The input layer size will be dictated by the number of inputs we have after preprocessing. You will see more on this soon. The middle layers can be either 1 or 2. It is best to choose the smallest number of neurons possible for a given problem to allow for generalization. If there are too many neurons, you will tend to get memorization of patterns. We will start with one hidden layer. The size of the first hidden layer is generally recommended as between one-half to three times the size of the input layer. If a second hidden layer is present, you may have between three and ten times the number of output neurons. The best way to determine optimum size is by trial and error.

Preprocessing Data

Neurons like to see data in a particular input range to be most effective. Presenting data that varies from 200 to 500 (as the S & P 500 has over the years) will not be useful, since the middle layer of neurons have a **Sigmoid Activation** function that squashes large inputs to either 0 or +1. In other words, you should choose data that fit a range that does not *saturate*, or overwhelm the network neurons. Choosing inputs from –1 to 1 or 0 to 1 is a good idea. The following lists are dimensions of preprocessing you should consider:

1. Smoothing the data for spikes.
2. Normalizing the input data for the 0 to 1 or –1 to 1 range.
3. Normalizing the expected values for the outputs to the 0 to 1 (sigmoid) range.
4. Presenting cyclic information.
5. Having the largest variability possible.
6. Avoiding too many data points at 0 and biasing appropriately.
7. Using first and second derivatives for feature detection.

These are a few ideas to present appropriate data for your neural network. The first, *smoothing*, is important to make sure that extraneous data points, which are not relevant to the generalization, are factored out or subdued. One important issue to consider when smoothing is that you don't introduce too much *time lag* between the smoothed data and the original. This is often the case when using moving averages of data.

There is often a cyclic component of data. If you are trying to make sales forecasts for your marketing department, you may notice a cyclic component to your raw data. The stock market is sometimes analyzed with cyclic measures. One method of presenting cyclic information is to take a Fourier transform of the input data. This will decompose the input discrete data series into a series of frequency spikes that measure the relevance of each frequency component. If the stock market indeed follows the so-called January effect, where prices typically make a run up, then you would expect a strong yearly component in the frequency spectrum. Mark Jurik suggests sampling data with intervals that catch different cycle periods, in his paper on neural network data preparation. (See references at the back of the book.)

It is often a good idea to highlight features for the network. By using differences in the stock market averages, rather than the raw levels, we are highlighting the changes. Similarly, it is often useful to present data that is related to the rate of change in the rate of change in data, which is the second derivative of the data set. In image processing, you can detect edges by accenting change with the function **(a-b)/(a+b)**, where a and b are adjacent pixel values. We will use a similar function to accent changes in the data.

It is important to pay attention to the number of values in the data set that are close to zero. Since the weight change law is proportional to the input value, then a close to zero input will mean that that weight will not participate in learning! To avoid such situations, you can add a constant bias to your data to move the data closer to 0.5, where the neurons respond very well.

A View of the Raw Data

Let's look at the raw data for the problem we want to solve. There are a few ways we can start preprocessing the data to reduce the number of inputs and enhance the variability of the data:

   ◆   Use Advances/Declines ratio instead of each value separately.

◆ Use New Highs/New Lows ratio instead of each value separately.

◆ Use Advancing Volume/Declining Volume ratio instead of each value separately.

We are left with the indicators:

1. S&P 500 High
2. S&P 500 Low
3. NYSE Advancing/Declining issues
4. NASDAQ Advancing/Declining issues
5. NYSE New Highs/New Lows
6. NASDAQ New Highs/New Lows
7. NYSE Total Volume
8. NYSE Advancing/Declining issues volume
9. NASDAQ Total Volume
10. NASDAQ Advancing/Declining issues volume
11. Three-Month Treasury Bill Yield
12. 30-Year Treasury Bond Yield
13. Gold
14. S & P 500 Closing price

Raw data for the period from January 4, 1980 to October 28, 1983 is taken as the training period, for a total of 200 weeks of data. The following 50 weeks are kept on reserve for a test period to see if the predictions are valid outside of the training interval. The last date of the test period is October 19, 1984. Let's look at the raw data now. (You get data on the disk available with this book that covers the period from January, 1980 to December, 1992.) In Figures 14.2 through 14.5, you will see a number of these indicators plotted over the training plus test intervals:

◆ Figure 14.2 shows you the S & P 500 stock index and the volume of shares traded on the NYSE (New York Stock Exchange).

◆ Figure 14.3 shows the inverse relationship between long-term interest rates as represented by the yield on the 30-year Treasury bond and inflation as measured by the price of gold.

◆ Figure 14.4 shows some breadth indicators on the NYSE, the number of advancing stocks/number of declining stocks, as well as the ratio of the volume of shares for advancing issues to the volume of shares for declining issues.

◆ Figure 14.5 shows the ratio of new highs to new lows on the NYSE, with the S&P 500 index as a reference.

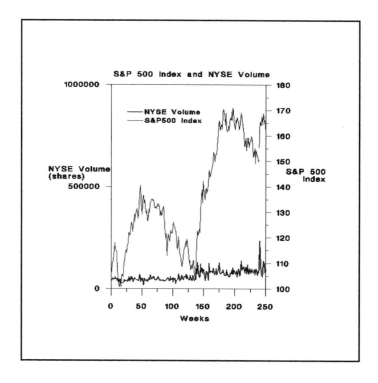

FIGURE 14.2   *The S&P 500 Index and the volume of shares on the NYSE versus for the training and test periods.*

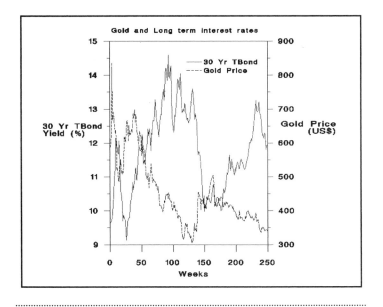

FIGURE 14.3 *Long-term interest rates and inflation as measured by the price of gold.*

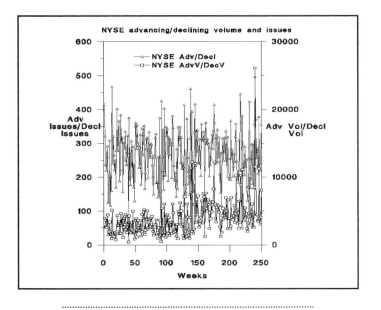

FIGURE 14.4 *Breadth indicators on the NYSE: Advancing /Declining issues and volume.*

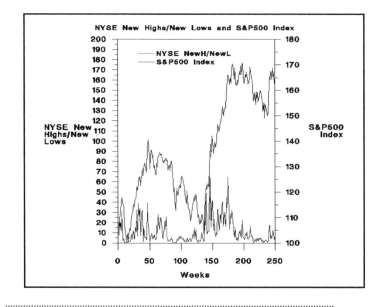

FIGURE 14.5 *Ratio of New Highs to New Lows on the NYSE, with the S & P 500 index.*

N O T E

A sample of a few lines looks like the following data. Note that the order of parameters is the same as listed above. The first value corresponds to the week ending January 4, 1980, and the last to the week ending January 25, 1980.

```
107.08  105.09  415.333  271.667  7.83333  8.66667  39130  3983.75  23480
    1826  12.11  9.64  588.0  106.52
111.16  108.89  321.667  224.333  24.8333  24.1667  52890  3643.75  28310
    2125.5  11.94  9.73  623.0  109.92
111.74  109.88  237.667  188  13.3333  21.1667  47150  2671.25  31880  1668
    11.90  9.80  835.0  111.07
114.45  112.36  220  171.667  19.8333  18.8333  47100  2541.25  25590  1562.88
    12.19  9.93  668.0  113.61
```

Taking Differences

The first step we take in preprocessing is taking differences in the data. The same lines are presented again after this processing is done.

```
0.000  0.000  0.000  0.000  0.000  0.000  0.000  0.000  0.000  0.000  0.000
    0.000  0.000  0.000
4.080  3.800  -93.666  -47.334  17.000  15.500  13760.000  -340.000  4830.000
    299.500  -0.170  0.090  35.000  3.400
0.580  0.990  -84.000  -36.333  -11.500  -3.000  -5740.000  -972.500  3570.000
    -457.500  -0.040  0.070  212.000  1.150  0.017
2.710  2.480  -17.667  -16.333  6.500  -2.333  -50.000  -130.000  -6290.000
    -105.120  0.290  0.130  -167.000  2.540
```

Note that the first line should be discarded, since it is the starting point.

NOTE

Normalizing the Range

We now have values that have a very large range. We would like to reduce the range by some method. We use the following function:

```
new value = (old value - Mean )/ standard deviation
```

This relates the distance from the mean for a value in a column as a fraction of the standard deviation for that column. A quick way to do this is with a spreadsheet, or, you could write a simple utility program to do this.

Note that we don't need to recalculate the mean and standard deviations each time we test new data; only after some time has elapsed and these values have shifted significantly. Make a note of the standard deviation and mean for each column of data, since you will need these values when you unnormalize data back to raw data form.

NOTE

The data after processing through this stage now looks like the following:

```
-0.094  -0.092  0.001  -0.005  -0.001  -0.004  -0.004  -0.002  -0.009  -0.006
    0.034  0.019  0.023  -0.092
0.692  0.628  -0.891  -0.526  1.729  1.761  0.297  -0.060  0.164  0.098  -0.405
    0.519  2.196  0.554
0.017  0.095  -0.799  -0.405  -1.172  -0.346  -0.130  -0.169  0.119  -0.167
    -0.069  0.407  13.182  0.126
```

Now that we have good variability, we notice that there are some values that are way out of the –1 to 1 or 0 to 1 range. We use the **Sigmoid**

Squashing function as a final step with a utility program that reads in a line, squashes each entry in the line, and writes a line to the output file. A sample utility program could look like the following file, called squash.cpp, Listing 14.1.

Listing 14.1   Utility program for squashing data, squash.cpp

```
// squash.cpp V. Rao, H. Rao
// This program takes 14 input fields and uses the sigmoid
    function on each field, writing results to
// an output file   Usage : squash inputfile outputfile

#include <stdio.h>
#include <iostream.h>
#include <stdlib.h>
#include <math.h>


inline float squash(float input)
// squashing function

{
if (input < -50)
    return 0.0;
else   if (input > 50)
        return 1.0;
    else return (float)(1/(1+exp(-(double)input)));

}

// the next function is needed for Turbo C++
// and Borland C++ to link in the appropriate
// functions for fscanf floating point formats:
static void force_fpf()
{
    float x, *y;
    y=&x;
    x=*y;
}

void main (int argc, char * argv[])
{

FILE * outfile, * infile;


float a[100];

int num_fields=14;

int i;
```

```
if (argc < 2)
    {
    cerr << "Usage: squash Inputfile Outputfile";
    exit(1);
    }
// open  files

infile= fopen(argv[1], "r");
outfile= fopen(argv[2], "w");

if ((infile == NULL) || (outfile == NULL))
    {
    cout << " Can't open a file\n";
    exit(1);
    }

while (!(feof(infile)))
    {
    // read next line
    for (i=0; i< num_fields; i++)
        {
        fscanf(infile,"%f",&a[i]);
        fprintf(outfile,"%f\t",(squash(a[i])));
        }

    fscanf(infile,"\n");
    fprintf(outfile,"\n");
    }

fclose(infile);
fclose(outfile);

}
```

After squashing, the data now looks like the following:

```
0.476517   0.477016   0.500250   0.498750   0.499750   0.499000
0.499000   0.499500   0.497750   0.498500   0.508499   0.504750
0.505750   0.477016
0.666412   0.652036   0.290904   0.371450   0.849284   0.853335
0.573709   0.485004   0.540908   0.524480   0.400112   0.626914
0.899890   0.635063
0.504250   0.523732   0.310239   0.400112   0.236494   0.414353
0.467546   0.457850   0.529715   0.458347   0.482757   0.600368
0.999998   0.531458
```

The data is well behaved and is in the 0 to 1 range. There are still a couple of more steps to take.

Adding an (a-b)/(a+b) term

As mentioned previously, it is useful to accent a change with the formula :

```
(current value - previous value) / (current value + previous value)
```

We do this on the original data, which gives us 14 additional columns of data as follows:

```
 0.018695   0.017759  -0.127091  -0.095431   0.520408   0.472082
 0.149533  -0.044576   0.093261   0.075794  -0.007069   0.004646
 0.028902   0.015709
 0.002602   0.004525  -0.150179  -0.088116  -0.301311  -0.066176
-0.057377  -0.153998   0.059312  -0.120601  -0.001678   0.003584
 0.145405   0.005204
 0.011981   0.011159  -0.038602  -0.045411   0.195980  -0.058335
-0.000531  -0.024940  -0.109448  -0.032536   0.012038   0.00658
-0.111111   0.011305
```

The Final Data Set (Almost)

You need to place the closing price as the last data field in a training field, so that the simulator knows that this is the target value. To combine the 14 columns obtained in the last step with the original 14 columns results in the following format. Note that the (a-b)/(a+b) terms appear first.

```
 0.018695   0.017759  -0.127091  -0.095431   0.520408   0.472082
 0.149533  -0.044576   0.093261   0.075794  -0.007069   0.004646
 0.028902   0.015709   0.476517   0.477016   0.500250   0.498750
 0.499750   0.499000   0.499000   0.499500   0.497750   0.498500
 0.508499   0.504750   0.505750   0.477016
 0.002602   0.004525  -0.150179  -0.088116  -0.301311  -0.066176
-0.057377  -0.153998   0.059312  -0.120601  -0.001678   0.003584
 0.145405   0.005204   0.666412   0.652036   0.290904   0.371450
 0.849284   0.853335   0.573709   0.485004   0.540908   0.524480
 0.400112   0.626914   0.899890   0.635063
 0.011981   0.011159  -0.038602  -0.045411   0.195980  -0.058335
-0.000531  -0.024940  -0.109448  -0.032536   0.012038   0.006589
-0.111111   0.011305   0.504250   0.523732   0.310239   0.400112
 0.236494   0.414353   0.467546   0.457850   0.529715   0.458347
 0.482757   0.600368   0.999998   0.531458
```

The final value in each row should be the target value, the weekly change in the S & P 500 index, which brings us to the final issue with preprocessing: time shifting.

Time Shifting the Data

So far, we have a data file that gives a number of indicators followed by the S & P 500 close for that week. (Actually, the change in the closing price from the last week to the current week.) In order to be useful for prediction, we need to train the network with next week's close based on this week's indicators. At the end of each week, we want to enter in all the data for this week and be able to predict next week's change in the S & P 500 index. We therefore must time shift the last column back by one entry. This can be done as a macro in some text editors like Lugaru's Epsilon, the emacs text editor, or you could write a simple utility program to do this. A sample of the final data set then, is as follows.

```
  0.018695   0.017759  -0.127091  -0.095431   0.520408   0.472082
  0.149533  -0.044576   0.093261   0.075794  -0.007069   0.004646
  0.028902   0.015709   0.476517   0.477016   0.500250   0.498750
  0.499750   0.499000   0.499000   0.499500   0.497750   0.498500
  0.508499   0.504750   0.505750   0.635063
  0.002602   0.004525  -0.150179  -0.088116  -0.301311  -0.066176
 -0.057377  -0.153998   0.059312  -0.120601  -0.001678   0.003584
  0.145405   0.005204   0.666412   0.652036   0.290904   0.371450
  0.849284   0.853335   0.573709   0.485004   0.540908   0.524480
  0.400112   0.626914   0.899890   0.531458
  0.011981   0.011159  -0.038602  -0.045411   0.195980  -0.058335
 -0.000531  -0.024940  -0.109448  -0.032536   0.012038   0.006589
 -0.111111   0.011305   0.504250   0.523732   0.310239   0.400112
  0.236494   0.414353   0.467546   0.457850   0.529715   0.458347
  0.482757   0.600368   0.999998   0.596523
```

Doing the Calculations for Preprocessing

You can use a spreadsheet for preprocessing the data, or you can write some simple utilities, similar to the squashing program (Listing 14.1) to do this. There are also graphing programs like CA-Cricket Graph, which allow you to do most of the manipulations mentioned. To summarize, we have taken the following steps in preprocessing.

1. Take delta of indicator from previous week, instead of absolute value.

2. For each delta, calculate the *z score*, or the distance of the delta from the mean of the column, normalized to a standard deviation of the column.

3. Squash the values from step 2 with the **Sigmoid** function to fall into the 0 to 1 range.

4. From raw data, calculate (a-b)/(a+b) to accent changes.

5. Combine data from steps 4 and 5.

6. Time shift the target value or answer that you will train to for prediction.

Storing Data in Different Files

You need to place the first 200 lines in the training.dat file (provided for you in the accompanying diskette) and the subsequent 50 lines of data in the another .dat file for use in testing. You will read more about this shortly. There is also more data than this provided on this diskette in raw and processed form for you to do further experiments.

Training and Testing

With the training data available, we set up a simulation. The number of inputs are 27, and the number of outputs is 1. A total of three layers are used with a middle layer size of 20. This number should be made as small as possible with acceptable results. The optimum sizes and number of layers can only be found by trial and error. The strategy is to use a high beta value for the initial runs, then load the weights from file for subsequent runs, and decrease beta. Also, after each run, you can look at the error from the training set and from the test set.

Using the Simulator to Calculate Error

You obtain the error for the test set by running the simulator in **Training** mode (you need to temporarily copy the test data with expected outputs to the training.dat file) for one cycle with weights loaded from the weights file. Since this is the last and only cycle, weights do not get modified, and you can get a reading of the average error. Refer to Chapter 13 for more information on the simulator's **Test** and **Training** modes. This approach has been taken with four runs of the simulator for 500, 500, 500, and 2,000 cycles. Table 14.1 summarizes the results along with the parameters used. Note that alpha and NF were set to zero for this particular problem.

TABLE 14.1 *Results of training the backpropagation simulator for predicting the change in next weeks S&P 500 index.*

Run#	Tolerance	Beta	Alpha	NF	max_cycles	cycles run	training set error	test set error
1	0.001	0.5	0	0	500	500	0.00308065	0.00909999
2	0.0001	0.5	0	0	500	500	0.00269243	0.0085594
3	0.0001	0.1	0	0	500	500	0.00114977	0.00666299
4	0.0001	0.1	0	0	2000	2000	0.000853455	0.00611207

N O T E

Note that in the last run, the test set error did not decrease much, whereas the training set error continued to make substantial progress. This means that memorization is starting to set in. It is important to monitor the test set(s) that you are using while you are training to make sure that good, generalized learning is occurring versus memorization or overfitting of the data.

To see the exact correlation, you can copy any period you'd like, with the expected value output fields deleted, to the test.dat file. Then you run the simulator in **Test** mode and get the output value from the simulator for each input vector. You can then compare this with the expected value in your training set or test set.

Now that you're done, you need to unnormalize the data back to get the answer in terms of the change in the S & P 500 index. What you've accomplished is a way in which you can get data from a financial newspaper, like *Barron's* or *Investor's Business Daily*, and feed the current week's data into your trained neural network to get a prediction of what the next week's S & P 500 index is likely to do.

Only the Beginning

This is only an example of what you can do with the simulator. You need to further analyze the data and present many, many more test cases to have a robust predictor. The prediction error obtained on the test data is 0.6%, which is quite good. The caveat though, is that this is only on a small sample of price behavior. More test data sets should be used on the simulator to monitor its effectiveness. A graph of the expected and predicted output for the test set and the training set is shown in Figure 14.6. The preprocessing steps shown in this chapter should serve as one example of the kinds of steps you can use. There are a vast variety of analysis and statistical methods that can be used in preprocessing. For applying fuzzy data, you can use a program like the fuzzifier program that was developed in Chapter 3 to preprocess some of the data.

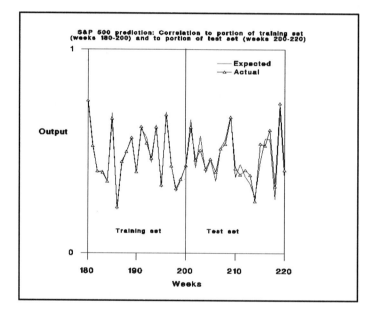

FIGURE 14.6 *Expected output versus actual for parts of the training and test data sets.*

There are many other experiments you can do from here on. The example chosen was in the field of financial forecasting. But you could certainly try the simulator on other problems like sales forecasting or perhaps even weather forecasting. The key to all of the applications though, is how you present data, and working through parameter selection by trial and error.

Summary

This chapter presented a neural network application in financial forecasting. Next week's change in the Standard & Poor's 500 stock index is the predicted output based on weekly changes in 14 indicators. Some examples of preprocessing of data for the network were shown as well as issues in training. The training of the backpropagation network took 3,500 cycles. At the end of the training period it was seen that memorization was beginning to take place, since the error in the test data did not decrease substantially, whereas the error in the training set did. It is important to monitor the error in the test data (without weight changes) while you are training to ensure that generalization ability is maintained.

This chapter's example in forecasting highlights the ease of use, and wide applicability of the backpropagation algorithm for large, complex problems and data sets.

Chapter 15

Application to Nonlinear Optimization

Introduction

Nonlinear optimization is an area of operations research, and efficient algorithms for some of the problems in this area are hard to find. An optimization problem has an **objective** function and a set of *constraints* on the variables. The problem is to find the values of the variables that lead to an optimum value for the **objective** function, while satisfying all the constraints. The **objective** function may be a **linear** function in the variables, or it may be a **nonlinear** function. For example, it could be a function expressing the total cost of a particular production plan, or a function giving the net profit from a group of products that share a given set of resources. The objective may be to find the minimum value for the **objective** function, if, for example, it represents cost, or to find the maximum value of a **profit** function. The resources shared by the products in their manufacturing are usually in limited supply or have some other restrictions on their availability. This consideration leads to the specification of the constraints for the problem.

Each constraint is usually in the form of an equation or an inequality. The left side of such an equation or inequality is an expression in the variables for the problem, and the right-hand side is a constant. The constraints are said to be linear or nonlinear depending on whether the expression on the left-hand side is a **linear** function or **nonlinear** function of the variables. A linear programming problem is an optimization problem with a linear objective function as well as a set of linear constraints. A nonlinear optimization problem has the **objective** function **nonlinear** and/or one or more of the constraints are nonlinear.

An example of a linear programming problem is the *blending* problem. An example of a blending problem is that of making different flavors of ice cream blending different ingredients, such as sugar, a variety of nuts, and so on, to produce different amounts of ice cream of many flavors. The objective in the problem is to find the amounts of individual flavors of ice cream to produce with given supplies of all the ingredients, so the total profit is maximized.

An example of a nonlinear optimization problem is the quadratic programming problem. The constraints are all linear but the **objective** function is a *quadratic form*. A quadratic form is an expression of two variables with 2 for the sum of the exponents of the two variables in each term.

Neural Networks for Optimization Problems

It is possible to construct a neural network to find the values of the variables that correspond to an optimum value of the **Objective** function of a problem. For example, the neural networks that use the Widrow-Hoff learning rule find the minimum value of the **Error** function using the least mean squared error. Neural networks such as the backpropagation network use the steepest descent method for this purpose and find a local minimum of the error, if not the global minimum. On the other hand, the Boltzman machine or the Cauchy machine uses statistical methods and probabilities and achieve success in finding the global minimum of an **Error** function. So we have an idea of how to go about using a neural network to find an optimum value of a function. The question remains as to how the constraints of an optimization problem should be treated in a neural network operation. A good example in answer to this question is the *traveling salesperson* problem. Let's discuss this example next.

Traveling Salesperson Problem

The traveling salesperson problem is a well-known problem in optimization. Its mathematical formulation is simple, and one can state a simple solution strategy also. Such a strategy is often impractical, and as yet, there is no efficient algorithm for this problem that consistently works in all instances. The traveling salesperson problem is one among the so called NP-complete problems, about which you will read more in what follows. That means that any algorithm for this problem is going to be impractical with certain examples. The neural network approach tends to give solutions with less computing time than other available algorithms for use on a digital computer. The problem is defined as follows. A traveling salesperson has a number of cities to visit. The sequence in which the salesperson visits different cities is called a *tour*. A tour should be such that every city on the list is visited once and only once, except that he returns to the city from which he starts. The goal is to find a tour that minimizes the total distance the salesperson travels, among all the tours that satisfy this criterion.

A simple strategy for this problem is to enumerate all feasible tours—a tour is feasible if it satisfies the criterion that every city is visited but once—to calculate the total distance for each tour, and to pick the tour with the smallest total distance. This simple strategy becomes impractical if the number of cities is large. For example, if there are 10 cities for the traveling salesperson to visit, there are 10! = 3,628,800 possible tours, where 10! denotes the factorial of 10—the product of all the integers from 1 to 10—and is the number of distinct permutations of the 10 cities. This number grows to over 6.2 billion with only 13 cities in the tour, and to over a trillion with 15 cities.

Formulation of Objective Function

The first consideration in the formulation of an optimization problem is the identification of the underlying variables and the type of values they can have. In a traveling salesperson problem, each city has to be visited once and only once, excepting the city started from. Suppose you take it for granted that the last leg of the tour is the travel between the last city visited and the city from which the tour starts, so that this part of the tour need not be explicitly included in the formulation. Then with n cities to be visited, the only information needed for any city is the position of that city in the order of visiting cities in the tour. This suggests that an ordered *n-tuple* is associated with each city with some element equal to 1 and the rest of the n – 1 elements equal to 0. In a neural network representation, this requires n neurons associated with one city. Only one of these n neurons corresponding to the position of the city, in the order of cities in the tour, fires or has output 1. Since there are n cities to be visited, you need n^2 neurons in the network. If these neurons are all arranged in a two-way array, you need a single 1 in each row and in each column of this array to indicate that each city is visited but only once.

Let x_{ij} be the variable to denote the fact that city i is the jth city visited in a tour. Then x_{ij} is the output of the jth neuron in the array of neurons corresponding to the ith city. You have n^2 such variables, and their values are binary, 0 or 1. In addition, only n of these variables should have value 1 in the solution. Furthermore, exactly one of the x's with the same first subscript (value of i) should have value 1. It is because the tour takes the salesperson from a given city to only one other city. Similarly, exactly one of the x's with the same second subscript (value of j) should have value 1. It is because the tour takes the salesperson to a given city from only one other city. These are the constraints in the problem. How do you then describe the tour? We take the city of start of the tour as city 1 in the array of cities. A tour can be given by the sequence x_{1a}, x_{ab}, x_{bc}, ..., x_{q1}, indicating that the cities visited in the tour in order starting at 1 are, a, b, c, ..., q and back to 1. Note that the sequence of subscripts a, b, ..., q is a permutation of 2, 3, ... n – 1.

With this set up, the distinct number of tours that satisfy the constraints is not n!, when there are n cities in the tour as given earlier. It is much less, namely, n!/2n. The reasons for this drop are discussed shortly. Thus when n is 10, the number of distinct feasible tours is 10!/20, which is

181,440. If n is 15, it is still over 43 billion, and it exceeds a trillion with 17 cities in the tour. Yet for practical purposes there is not much comfort knowing that for the case of 13 cities, 13! is over 6.2 billion and 13!/26 is only 239.5 million—it still is a tough combinational problem.

The cost of the tour is the total distance traveled, and it is to be minimized. The total distance traveled in the tour is the sum of the distances from one city to the next. The **objective** function has one term that corresponds to the total distance traveled in the tour. The other terms, one for each constraint, in the **objective** function are expressions, each attaining a minimum value if and only if the corresponding constraint is satisfied. The **objective** function then takes the following form. Hopfield and Tank formulated the problem as one of minimizing energy. It is, thus, customary to refer to the value of the **objective** function of this problem, while using a neural network for its solution, as the energy level of the network. The goal is to minimize this energy level.

We will denote this energy by the letter E. In formulating the equation for E, one uses constant parameters A_1, A_2, A_3, and A_4 as coefficients in different terms of the expression on the right-hand side of the equation. The equation that defines E is given as follows. Note that the notation in this equation includes d_{ij} for the distance from city i to city j.

$$E = A_1 \sum_i \sum_k \sum_{j \neq k} x_{ik} x_{ij} + A_2 \sum_i \sum_k \sum_{j \neq k} x_{ki} x_{ji} + A_3 [(\sum_i \sum_k x_{ik})$$
$$- n]^2 + A_4 \sum_k \sum_{j \neq k} \sum_i d_{kj} x_{ki}(x_{j,i+1} + x_{j,i-1})$$

Our first observation at this point is that E is a **nonlinear** function of the x's, as you have quadratic terms in it. So this formulation of the traveling salesperson problem renders it a nonlinear optimization problem.

All the summations indicated by the occurrences of the summation symbol $\sum$, range from 1 to n for the values of their respective indices. This means that the same summand such as $x_{12}x_{33}$ also as $x_{13}x_{32}$, appears twice with only the factors interchanged in their order of occurrence in the summand. For this reason, many authors use an additional factor of 1/2 for each term in the expression for E. However, when you minimize a quantity z with a given set of values for the variables in the expression for z, the same values for these variables minimize any whole or fractional multiple of z as well.

The third summation in the first term is over the index j, from 1 to n, but excluding whatever value k has. this prevents you from using something like $x_{12}x_{12}$. Thus the first term is an abbreviation for the sum of $n^2(n-1)$ terms with no two factors in a summand equal. This term is included to correspond to the constraint that no more than one neuron in the same row can output a 1. You thus get 0 for this term with a valid solution. So is the second term in the right-hand side of the equation for E. Note also that for any value of the index i, x_{ii} has value 0, since you are not making a move like, from city i to the same city i in any of the tours you consider as a solution to this problem. The third term in the expression for E has a minimum value of 0, which is attained if and only if exactly n of the n^2 x's have value 1 and the rest 0.

The last term is the one expressing the goal of finding a tour with the least total distance traveled, indicating the shortest tour among all possible tours for the traveling salesperson. Another important issue about the values of the subscripts on the right-hand side of the equation for E is, what happens to $i + 1$, for example, when i is already equal to n, and to $i-1$, when i is equal to 1. The $i + 1$ and $i - 1$ seem like impossible values, being out of their allowed range from 1 to n. The trick is to replace these values with their moduli with respect to n. This means, that the value $n + 1$ is replaced with 1, and the value 0 is replaced with n in the situations just described.

Modular values are obtained as follows. If we want, say 13 modulo 5, we subtract 5 as many times as possible from 13, until the remainder is a value between 0 and 4, 4 being $5 - 1$. Since we can subtract 5 twice from 13 to get a remainder of 3, which is between 0 and 4, 3 is the value of 13 modulo 5. Thus $(n + 3)$ modulo n is 3, as previously noted. Another way of looking at these results is that 3 is 13 modulo 5 because, if you subtract 3 from 13, you get a number divisible by 5, or which has 5 as a factor. Subtracting 3 from $n + 3$ gives you n, which has n as a factor. So 3 is $(n + 3)$ modulo n. In the case of -1, by subtracting $(n - 1)$ from it, we get -n, which can be divided by n getting -1. So $(n - 1)$ is the value of (-1) modulo n.

Example of a Traveling Salesperson Problem for Hand Calculation

Suppose there are four cities in the tour. Call these cities, C_1, C_2, C_3, and C_4. Let the matrix of distances be the following matrix D.

$$D = \begin{matrix} 0 & 10 & 14 & 7 \\ 10 & 0 & 6 & 12 \\ 14 & 6 & 0 & 9 \\ 7 & 12 & 9 & 0 \end{matrix}$$

From our earlier discussion on the number of valid and distinct tours, we infer that there are just three such tours. Since it is such a small number, we can afford to enumerate the three tours, find the energy values associated with them, and pick the tour that has the smallest energy level among the three. The three tours are:

◆ **Tour 1.** $1 - 2 - 3 - 4 - 1$ In this tour city 2 is visited first, followed by city 3, from where the salesperson goes to city 4, and then returns to city 1. For city 2, the corresponding ordered array is (1, 0, 0, 0), because city 2 is the first in this permutation of cities. Then $x_{21} = 1$, $x_{22} = 0$, $x_{23} = 0$, $x_{24} = 0$. Also (0, 1, 0, 0), (0, 0, 1, 0), and (0, 0, 0, 1) correspond to cities 3, 4, and 1, respectively. The total distance of the tour is $d_{12} + d_{23} + d_{34} + d_{41} = 10 + 6 + 9 + 7 = 32$.

◆ **Tour 2.** $1 - 3 - 4 - 2 - 1$

◆ **Tour 3.** $1 - 4 - 2 - 3 - 1$

There seems to be some discrepancy here. If there is one, we need an explanation. The discrepancy is that we can find many more tours that should be valid because no city is visited more than once. You may say they are distinct from the three previously listed. Some of these additional tours are:

◆ **Tour 4.** $1 - 2 - 4 - 3 - 1$

◆ **Tour 5.** $3 - 2 - 4 - 1 - 3$

◆ **Tour 6.** $2 - 1 - 4 - 3 - 2$

There is no doubt that these three tours are distinct from the first set of three tours. And in each of these three tours, every city is visited exactly once, as required in the problem. So they are valid tours as well. Why did our formula give us 3 for the value of the number of possible valid tours, while we are able to find 6?

The answer lies in the fact that if two valid tours are symmetric and have the same energy level, because they have the same value for the total distance traveled in the tour, one of them is in a sense redundant, or one of them can be stated to be *degenerate*, using the terminology common to this context. As long as they are valid and give the same total distance, the two tours are not individually interesting, and any one of the two is enough to have. By simple inspection, you find the total distances for the six listed tours. They are 32 for tour 1, 32 also for tour 6, 45 for each of tours 2 and 4, and 39 for each of tours 3 and 5. Notice also that tour 6 is not very different from tour 1. Instead of starting at city 1 as in tour 1, if you start at city 2 and follow tour 1 from there in reverse order of visiting cities, you get tour 6. Therefore, the distance covered is the same for both these tours. You can find similar relationships for other pairs of tours that have the same total distance for the tour. Either by reversing the order of cities in a tour, or by making a circular permutation of the order of the cities in a tour, you get another tour with the same total distance. This way you can find tours. Thus, only three distinct total distances are possible, and among them 32 is the lowest. The tour 1 – 2 – 3 – 4 – 1 is the shortest and is an optimum solution for this traveling salesperson problem. There is an alternative optimum solution, namely tour 6, with 32 for the total length of the tour. The problem is to find an optimal tour, and not to find all optimal tours if more than one exist. That is the reason only three distinct tours are suggested by the formula for the number of distinct valid tours in this case.

The formula we used to get the number of valid and distinct tours to be 3 is based on the elimination of such degeneracy. To clarify all this discussion, you should determine the energy levels of all your six tours identified earlier, hoping to find pairs of tours with identical energy levels.

Note that the first term on the right-hand side of the equation results in 0 for a valid tour, as this term is to ensure there is no more than a single 1 in each row. That is, in any summand in the first term, at least one of the factors, x_{ik} or x_{ij}, where $k \neq j$ has to be 0 for a valid tour. So all those summands are 0, making the first term itself 0. Similarly, the second term is 0 for a valid tour, because in any summand at least one of the factors, x_{ki} or x_{ji}, where $k \neq j$ has to be 0 for a valid tour. In all, exactly 4 of the 16 x's are each 1, making the total of the x's 4. This causes the third term to be 0 for a valid tour. These observations make it clear that it does not matter for valid tours, what values are assigned to the parameters A_1, A_2, and A_3. Assigning large values for these parameters would cause the energy levels, for tours that are not valid, to be much larger than the energy levels for valid tours. Thereby, these tours become not attractive for the solution of the traveling salesperson problem. Let us use the value 1/2 for the parameter A_4.

Let us demonstrate the calculation of the value for the last term in the equation for E, in the case of tour 1. Recall that the needed equation is

$$E = A_1 \sum_i \sum_k \sum_{j \neq k} x_{ik} x_{ij} + A_2 \sum_i \sum_k \sum_{j \neq k} x_{ki} x_{ji} + A_3[(\sum_i \sum_k x_{ik})$$
$$- n]^2 + A_4 \sum_k \sum_{j \neq k} \sum_i d_{kj} x_{ki}(x_{j,i+1} + x_{j,i-1})$$

The last term expands as given in the following calculation:

$A_4\{ d_{12}[x_{12}(x_{23} + x_{21}) + x_{13}(x_{24} + x_{22}) + x_{14}(x_{21} + x_{23})] +$
$d_{13}[x_{12}(x_{33} + x_{31}) + x_{13}(x_{34} + x_{32}) + x_{14}(x_{31} + x_{33})] +$
$d_{14}[x_{12}(x_{43} + x_{41}) + x_{13}(x_{44} + x_{42}) + x_{14}(x_{41} + x_{43})] +$
$d_{21}[x_{21}(x_{12} + x_{14}) + x_{23}(x_{14} + x_{12}) + x_{24}(x_{11} + x_{13})] +$
$d_{23}[x_{21}(x_{32} + x_{34}) + x_{23}(x_{34} + x_{32}) + x_{24}(x_{31} + x_{33})] +$
$d_{24}[x_{21}(x_{42} + x_{44}) + x_{23}(x_{44} + x_{42}) + x_{24}(x_{41} + x_{43})] +$
$d_{31}[x_{31}(x_{12} + x_{14}) + x_{32}(x_{13} + x_{11}) + x_{34}(x_{11} + x_{13})] +$
$d_{32}[x_{31}(x_{22} + x_{24}) + x_{32}(x_{23} + x_{21}) + x_{34}(x_{21} + x_{23})] +$
$d_{34}[x_{31}(x_{42} + x_{44}) + x_{32}(x_{43} + x_{41}) + x_{34}(x_{41} + x_{43})] +$
$d_{41}[x_{41}(x_{12} + x_{14}) + x_{42}(x_{13} + x_{11}) + x_{43}(x_{14} + x_{12})] +$
$d_{42}[x_{41}(x_{22} + x_{24}) + x_{42}(x_{23} + x_{21}) + x_{43}(x_{24} + x_{22})] +$
$d_{43}[x_{41}(x_{32} + x_{34}) + x_{42}(x_{33} + x_{31}) + x_{43}(x_{34} + x_{32})] \}$

When the respective values are substituted for the tour $1 - 2 - 3 - 4 - 1$, the above calculation becomes:

$1/2\{10[0(0 + 1) + 0(0 + 0) + 1(1 + 0)] + 14[0(0 + 0) +$
$\quad 0(0 + 1) + 1(0 + 0)] +$
$7[0(1 + 0) + 0(0 + 0) + 1(0 + 1)] + 10[1(0 + 1) + 0(1 + 0)$
$\quad + 0(0 + 0)] +$
$6[1(1 + 0) + 0(0 + 1) + 0(0 + 0)] + 12[1(0 + 0) + 0(0 + 0)$
$\quad + 0(0 + 1)] +$
$14[0(0 + 1) + 1(0 + 0) + 0(0 + 0)] + 6[0(0 + 0) + 1(0 + 1)$
$\quad + 0(1 + 0)] +$
$9[0(0 + 0) + 1(1 + 0) + 0(0 + 1)] + 7[0(0 + 1) + 0(0 + 0)$
$\quad + 1(1 + 0)] +$
$12[0(0 + 0) + 0(0 + 1) + 1(0 + 0)] + 9[0(1 + 0) + 0(0 +$
$\quad 0) + 1(0 + 1)]\}$
$= 1/2(10 + 0 + 7 + 10 + 6 + 0 + 0 + 6 + 9 + 7 + 0 + 9)$

```
= 1/2(64)
= 32
```

Table 15.1 contains the values we get for the fourth term on the right-hand side of the equation, and for E, with the six listed tours.

TABLE 15.1   *Energy levels for six of the valid tours.*

tour #	no-zero x's	value for the last term	energy level	comment
1	$x_{14}, x_{21}, x_{32}, x_{43}$	32	32	1 – 2 – 3 – 4 – 1 tour
2	$x_{14}, x_{23}, x_{31}, x_{42}$	45	45	1 – 3 – 4 – 2 – 1 tour
3	$x_{14}, x_{22}, x_{33}, x_{41}$	39	39	1 – 4 – 2 – 3 – 1 tour
4	$x_{14}, x_{21}, x_{33}, x_{42}$	45	45	1 – 2 – 4 – 3 – 1 tour
5	$x_{13}, x_{21}, x_{34}, x_{42}$	39	39	3 – 2 – 4 – 1 – 3 tour
6	$x_{11}, x_{24}, x_{33}, x_{42}$	32	32	2 – 1 – 4 – 3 – 2 tour

Neural Network for Traveling Salesperson Problem

Hopfield and Tank used a neural network to solve a traveling salesperson problem. The solution you get may not be optimal in certain instances. But by and large you may get a solution close to an optimal solution. One cannot find for the traveling salesperson problem a consistently efficient method of solution, as it has been proven to belong to a set of problems called *NP*-complete problems. NP-complete problems have the distinction that there is no known algorithm that is efficient and practical, and there is no likelihood that such an algorithm will be developed in the future. This is a caveat to keep in mind when using a neural network to solve a traveling salesperson problem.

Network Choice and Lay Out

We will describe the use of a Hopfield network to attempt to solve the traveling salesperson problem. There are n^2 neurons in the network arranged in a two-dimensional array of n neurons per row and n per column. The network is fully connected. The connections in the network in each row and in each column are lateral connections. The layout of the neurons in the network with their connections is shown in Figure 15.1 for the case of three cities, for illustration. To avoid cluttering, the connections between diagonally opposite neurons are not shown.

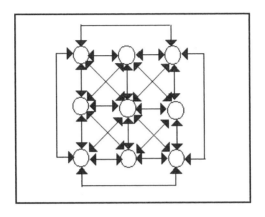

FIGURE. 15.1   *Layout of a Hopfield network for the Traveling Salesperson Problem.*

The most important task on hand then is finding an appropriate connection weight matrix. It is constructed taking into account that nonvalid tours should be prevented and valid tours should be preferred. One of the considerations in this regard is, for example, no two neurons in the same row (column) should fire in the same cycle of operation of the network. As a consequence, the lateral connections should be for inhibition and not for excitation.

In this context, the **Kronecker Delta** function is used to facilitate simple notation. The **Kronecker Delta** function has two arguments, which are given usually as subscripts of the symbol δ. By definition of δ_{ik} has value 1 if i = k, and 0 if i ≠ k. That is, the two subscripts should agree for the Kronecker delta to have value 1. Otherwise, its value is 0.

We refer to the neurons with two subscripts, one for the city it refers to, and the other for the order of that city in the tour. Therefore, an element of the weight matrix for a connection between two neurons needs to have four subscripts, with a comma after two of the subscripts. An example is $w_{ik,lj}$

referring to the weight on the connection between the (ik) neuron and the (lj) neuron. The value of this weight is set as follows:

$$w_{ik,lj} = -A_1\delta_{il}(1-\delta_{kj}) - A_2\delta_{kj}(1-\delta_{il}) - A3 - A_4\ d_{il}(\delta_{j,\ k+1} + \delta_{j,k-1})$$

Here the negative signs indicate inhibition through the lateral connections in a row or column. The $-A_3$ is a term for global inhibition.

Inputs

The inputs to the network are chosen arbitrarily. If as a consequence of the choice of the inputs, the activations work out to give outputs that add up to the number of cities, an initial solution for the problem, a legal tour, will result. A problem may also arise that the network will get stuck at a local minimum. To avoid such an occurrence, random noise can be added. Usually you take as the input at each neuron, a constant times the number of cities and adjust this adding a random number, which can be different for different neurons.

Activations, Outputs, and Their Updating

We denote the activation of the neuron in the ith row and jth column by a_{ij}, and the output is denoted by x_{ij}. A time constant τ, and a gain λ are used as well. A constant m is another parameter used. Also, Δt denotes the increment in time, from one cycle to the next. Keep in mind that the index for the summation Σ ranges from 1 to n, the number of cities. Excluded values of the index are shown by the use of the symbol $\neq$.

The change in the activation is then given by Δa_{ij}, where

$$\Delta a_{ij} = \Delta t\ (Term_1 + Term_2 + Term_3 + Term_4 + Term_5)$$
$$Term_1 = -\ a_{ij}/\tau$$
$$Term_2 = -\ A_1 \sum_{k \neq j} x_{ik}$$
$$Term_3 = -\ A_2 \sum_{k \neq i} x_{kj}$$
$$Term_4 = -\ A_3(\sum_i \sum_k x_{ik} - m)$$
$$Term_5 = -\ A_4 \sum_{k \neq i} d_{ik}(x_{k,j+1} + x_{k,j-1})$$

To update the activation of the neuron in the ith row and jth column, you take

$$a_{ijnew} = a_{ijold} + \Delta a_{ij}$$

The output of a neuron in the ith row and jth column is calculated from

$$x_{ij} = (1 + \tanh(\lambda a_{ij}))/2$$

The function used here is the hyperbolic tangent function. λ is the gain parameter mentioned earlier. The output of each neuron is calculated after updating the activation. Ideally, you want to get the outputs as 0's and 1's, preferably a single one for each row and each column, to represent a tour that satisfies the conditions of the problem. But the hyperbolic tangent function gives a real number, and you have to settle for a close enough value to 1 or 0.

Performance of Hopfield Network

Hopfield and Tank Example

The Hopfield network's use in solving the traveling salesperson problem is a pioneering effort in the use of neural network approach for this problem. Hopfield's example is for a problem with 10 cities. The parameters used were, $A_1 = 500$, $A_2 = 500$, $A_3 = 200$, $A_4 = 500$, $\tau = 1$, $\lambda = 50$, and m = 15. A good solution is expected, if not the best solution, because of the possibility of settling down at a local minimum. An annealing process could be considered to move out of any local minimum. As was mentioned before, the traveling salesperson problem is one of those problems for which a single approach cannot be found that will be successful in all cases.

Comment on Parameters

There isn't very much guidance as to how to choose the parameters in general for the use of the Hopfield network to solve the traveling salesperson problem. Since the **energy** function is minimized with the shortest distance for a tour in which each city is visited exactly once and the salesperson returns to the first city, it makes sense to try the neural network approach to this problem.

C++ Implementation of Hopfield Network for the Traveling Salesperson Problem

We present a C++ program for the Hopfield network operation for the traveling salesperson problem. The header file is in the Listing 15.1, and the source file is in the Listing 15.2. A **tsneuron** class is declared for a neuron and a **Network** class for the network. The **Network** class is declared a friend in the **tsneuron** class. The program follows the procedure described for setting inputs, connection weights, and updating.

Listing 15.1 **Header file for the C++ program for the Hopfield network for the traveling salesperson problem**

```
//trvslsmn.h   V. Rao,  H. Rao
#include <iostream.h>
#include <stdlib.h>
#include <math.h>
#include <stdio.h>

#define MXSIZ 11


class tsneuron
    {
    protected:
    int cit,ord;
    float output;
    float activation;
    friend class network;

    public:
    tsneuron() { };
    void getnrn(int,int);
    };


class network
```

```
{
public:
int  citnbr;
float pra,prb,prc,prd,totout,distnce;

tsneuron (tnrn)[MXSIZ][MXSIZ];
int dist[MXSIZ][MXSIZ];
int tourcity[MXSIZ];
float outs[MXSIZ][MXSIZ];
float acts[MXSIZ][MXSIZ];
float mtrx[MXSIZ][MXSIZ];
float citouts[MXSIZ];
float ordouts[MXSIZ];

network() { };
void getnwk(int,float,float,float,float);
void getdist(int);
void findtour();
void asgninpt(float *);
void calcdist();
void iterate(int,int,float,float,float);
void getacts(int,float,float);
void getouts(float);

//print functions

void prdist();
void prmtrx(int);
void prtour();
void practs();
void prouts();
};
```

Source File for Hopfield Network for Traveling Salesperson Problem

The following listing gives the source code for the C++ program for the Hopfield network for traveling salesperson problem. The user is prompted to input the number of cities and the maximum number of iterations for the operation of the network.

The parameters a, b, c, d declared in the function main correspond to A_1, A_2, A_3, and A_4, respectively. These and the parameters tau, lambda, and nprm are all given values in the declaration statements in the function main. If you change these parameter values or change the number of cities in the traveling salesperson problem, the program will compile but may not work as you'd like.

Listing 15.2   Source file for the C++ program for the Hopfield network for the traveling salesperson problem

```
//trvslsmn.cpp V. Rao,  H. Rao
#include "trvslsmn.h"
#include <stdlib.h>
#include <time.h>


//generate random noise

int randomnum(int maxval)
{
// random number generator
// will return an integer up to maxval

return rand() % maxval;
}



//Kronecker delta function

int krondelt(int i,int j)
    {
    int k;
    k= ((i == j) ? (1):(0));
    return k;
    };

void tsneuron::getnrn(int i,int j)
    {
    cit = i;
```

```
        ord = j;
        output = 0.0;
        activation = 0.0;
        };

//distances between cities

void network::getdist(int k)
    {
    citnbr = k;
    int i,j;
    cout<<"\n";

    for(i=0;i<citnbr;++i)
        {
        dist[i][i]=0;
        for(j=i+1;j<citnbr;++j)
            {
            cout<<"\ntype distance (integer) from city "<<
            i<<" to city "<<j<<"\n";
            cin>>dist[i][j];
            }
        cout<<"\n";
        }

    for(i=0;i<citnbr;++i)
        {
        for(j=0;j<i;++j)
            {
            dist[i][j] = dist[j][i];
            }
        }
    prdist();
    cout<<"\n";
    }


//print distance matrix
```

```
void network::prdist()
   {
   int i,j;
   cout<<"\n";

   for(i=0;i<citnbr;++i)
      {
      for(j=0;j<citnbr;++j)
         {
         cout<<dist[i][j]<<"   ";
         }
      cout<<"\n";
      }
   }


//set up network

void network::getnwk(int citynum,float a,float b,float
   c,float d)
   {
   int i,j,k,l,t1,t2,t3,t4,t5,t6;
   int p,q;
   citnbr = citynum;
   pra = a;
   prb = b;
   prc = c;
   prd = d;
   getdist(citnbr);

   for(i=0;i<citnbr;++i)
      {
      for(j=0;j<citnbr;++j)
         {
         tnrn[i][j].getnrn(i,j);
         }
      }
```

```
//find weight matrix

for(i=0;i<citnbr;++i)
    {
    for(j=0;j<citnbr;++j)
        {
        t1 = j + i*citnbr;
        for(k=0;k<citnbr;++k)
            {
            for(l=0;l<citnbr;++l)
                {
                t2 = l + k*citnbr;
                t3 = krondelt(i,k);
                t4 = krondelt(j,l);
                p = ((j == citnbr-1) ? (0) : (j+1));
                q = ((j == 0) ? (citnbr-1) : (j-1));
                t5 = krondelt(l,p);
                t6 = krondelt(l,q);
                mtrx[t1][t2] =
                -a*t3*(1-t4) -b*t4*(1-t3)
                -c -d*t3*dist[i][j]*(t5+t6);
                }
            }
        }
    prmtrx(citnbr);
    }

//print weight matrix

void network::prmtrx(int k)
    {
    int i,j,nbrsq;
    nbrsq = k*k;

    for(i=0;i<nbrsq;++i)
        {
```

```
        for(j=0;j<nbrsq;++j)
            {
            if(j%k == 0)
                {
                cout<<"\n";
                }
            cout<<mtrx[i][j]<<"  ";
            }
        cout<<"\n";
        }
    }

//present input to network

void network::asgninpt(float *ip)
    {
    int i,j,k,l,t1,t2;

    for(i=0;i<citnbr;++i)
        {
        for(j =0;j<citnbr;++j)
            {
            acts[i][j] = 0.0;
            }
        }

    //find initial activations
    for(i=0;i<citnbr;++i)
        {
        for(j =0;j<citnbr;++j)
            {
            t1 = j + i*citnbr;
            for(k=0;k<citnbr;++k)
                {
                for(l=0;l<citnbr;++l)
                    {
                    t2 = l + k*citnbr;
                    acts[i][j] +=
```

```
                    mtrx[t1][t2]*ip[t1];
                }
            }
        acts[i][j] += prc*citnbr;
        }
    }


//print activations

practs();
}



//find activations

void network::getacts(int nprm,float dlt,float tau)
    {
    int i,j,k,p,q;
    float r1, r2, r3, r4,r5;
    r3 = totout - nprm ;

    for(i=0;i<citnbr;++i)
        {
        r4 = 0.0;
        p = ((i == citnbr-1) ? (0) : (i+1));
        q = ((i == 0) ? (citnbr-1) : (i-1));
        for(j=0;j<citnbr;++j)
            {
            r1 = citouts[i] - outs[i][j];
            r2 = ordouts[i] - outs[i][j];
            for(k=0;k<citnbr;++k)
                {
                r4 += dist[i][k] *
                (outs[k][p] + outs[k][q]);
                }
            r5 = dlt*(-acts[i][j]/tau -
            pra*r1 -prb*r2 -prc*r3 -prd*r4);
            acts[i][j] += r5;
```

```
                }
            }
        }


//find outputs and totals for rows and columns

void network::getouts(float la)
    {
    float b1,b2,b3,b4;
    int i,j;
    totout = 0.0;

    for(i=0;i<citnbr;++i)
        {
        citouts[i] = 0.0;
        for(j=0;j<citnbr;++j)
            {
            b1 = la*acts[i][j];
            b4 = b1/500.0;
            b2 = exp(b4);
            b3 = exp(-b4);
            outs[i][j] = (1.0+(b2-b3)/(b2+b3))/2.0;
            citouts[i] += outs[i][j];};
            totout += citouts[i];
            }
        for(j=0;j<citnbr;++j)
            {
            ordouts[j] = 0.0;
            for(i=0;i<citnbr;++i)
                {
                ordouts[j] += outs[i][j];
                }
            }
        }

//find tour

void network::findtour()
```

```
{
int i,j,k,tag[MXSIZ][MXSIZ];
float tmp;
for (i=0;i<citnbr;++i)
    {
    for(j=0;j<citnbr;++j)
        {
        tag[i][j] = 0;
        }
    }
    for (i=0;i<citnbr;++i)
        {
        tmp = -10.0;
        for(j=0;j<citnbr;++j)
            {
            for(k=0;k<citnbr;++k)
                {
                if((outs[i][k] >=tmp)&&
                (tag[i][k] ==0))
                    tmp = outs[i][k];
                }
            if((outs[i][j] ==tmp)&&
            (tag[i][j]==0))
                {
                tourcity[i] =j;
                cout<<"\ntourcity "<<i
                <<" tour order "<<j<<"\n";
                for(k=0;k<citnbr;++k)
                    {
                    tag[i][k] = 1;
                    tag[k][j] = 1;
                    }
                }
            }
        }
    }
```

```
//print outputs

void network::prouts()
    {
    int i,j;
    cout<<"\nthe outputs\n";
    for(i=0;i<citnbr;++i)
        {
        for(j=0;j<citnbr;++j)
            {
            cout<<outs[i][j]<<"  ";
            }
        cout<<"\n";
        }
    }


//calculate total distance for tour

void network::calcdist()
    {
    int i, k, l;
    distnce = 0.0;

    for(i=0;i<citnbr;++i)
        {
        k = tourcity[i];
        l = ((i == citnbr-1 ) ? (tourcity[0]):(tourcity[i+1]));
        distnce += dist[k][l];
        }
    cout<<"\n distance of tour is : "<<distnce<<"\n";
    }


// print tour

void network::prtour()
    {
    int i;
    cout<<"\n the tour :\n";
```

```
for(i=0;i<citnbr;++i)
    {
    cout<<tourcity[i]<<"  ";
    cout<<"\n";
    }
}


//print activations

void network::practs()
    {
    int i,j;
    cout<<"\n the activations:\n";
    for(i=0;i<citnbr;++i)
        {
        for(j=0;j<citnbr;++j)
            {
            cout<<acts[i][j]<<"  ";
            }
        cout<<"\n";
        }
    }


//iterate specified number of times

void network::iterate(int nit,int nprm,float dlt,float
    tau,float la)
    {
    int k;

    for(k=1;k<=nit;++k)
        {
        getacts(nprm,dlt,tau);
        getouts(la);
        }
    cout<<"\n";
    practs();
```

```
    cout<<"\n";
    prouts();
    prtour();
    calcdist();
    cout<<"\n";
    }


void main()
    {

//numit - #iterations; n - #cities; u-initial input; nprm -
// parameter n'
//dt - delta t;
// ----------------------------------
// parameters to be tuned are here:
    int u=1;
    int nprm=10;
    float a=40.0;
    float b=40.0;
    float c=30.0;
    float d=60.0;
    float dt=0.01;
    float tau=1.0;
    float lambda=3.0;
//-----------------------------------
    int i,n2;
    int numit=100;
    int n=4;
    float input_vector[MXSIZ*MXSIZ];

    srand ((unsigned)time(NULL));

    cout<<"\nPlease type number of cities, number of itera-
    tions\n";

    cin>>n>>numit;
    cout<<"\n";
    if (n>MXSIZ)
```

```
        {
        cout << "choose a smaller n value\n";
        exit(1);
        }
    n2 = n*n;
    for(i=0;i<n2;++i)
        {
        if(i%n == 0)cout<<"\n";
        input_vector[i] =(u + (float)(random-
    num(100)/100.0))/20.0;
        cout<<input_vector[i]<<" ";
        }


//create network and operate

    network *tntwk = new network;
    if (tntwk==0)
        {
        cout << "not enough memory\n";
        exit(1);
        }
    tntwk->getnwk(n,a,b,c,d);
    tntwk->asgninpt(input_vector);
    tntwk->getouts(lambda);
    tntwk->prouts();
    tntwk->iterate(numit,nprm,dt,tau,lambda);
    tntwk->findtour();
    tntwk->prtour();
    tntwk->calcdist();
    cout<<"\n";
}
```

Output from Your C++ Program for the Traveling Salesperson Problem

The program is run for two cases, for illustration. The first run is for a problem with three cities. The second one is for a four-city problem. The solution you get for the three-city problem is not the one in natural order. In the case of the four-city problem, the tour in the natural order is the output. It of course is a tour of shortest distance, with the distance matrix given. By the way, the cities are numbered from 0 to n - 1. The same parameter values are used in the two runs. The number of cities, and consequently, the matrix of distance were different. In both cases, the number of iterations asked for is 30.

The program gives you the order in the tour for each city. For example, if it says tourcity 1 tour order 2, that means the second city is the third city visited. Your tour orders are also with values from 0 to n – 1, like the cities.The example here means the tour is from city 0 to city 2 and then to city 1 from which you return to city 0. It can be shown as $0 \rightarrow 2 \rightarrow 1 \rightarrow 0$, for a total distance of 31, in this case.

The computer output is in boldface, and user input is normal, as you have seen before.

Output for a Three-City Problem

Please type number of cities, number of iterations
3   30

0.001027 0.000385 0.033491   `//input vector—there are 9 neu-`
                                        `rons in the network`
0.003294 0.035586 0.021728
0.053679 0.019547 0.07006

type distance from city 0 to city 1
10

type distance from city 0 to city 2
14

type distance from city 1 to city 2
7

0 10 14   `// the matrix of distances`
10 0 7
14 7 0

-30 -70 -70 -70

//the 9 x 9 matrix of weights, there
are nine neurons in the network

-30 -30 -70 -30
-30

-670 -30 -670 -30
-70 -30 -30 -70
-30

-910 -910 -30 -30
-30 -70 -30 -30
-70

-70 -30 -30 -30
-670 -670 -70 -30
-30

-30 -70 -30 -70
-30 -70 -30 -70
-30

-30 -30 -70 -490
-490 -30 -30 -30
-70

-70 -30 -30 -70
-30 -30 -30 -910
-910

-30 -70 -30 -30
-70 -30 -490 -30
-490

-30 -30 -70 -30
-30 -70 -70 -70
-30

 the activations:
-0.441587 -0.628157 -70.665145
-5.368961 -15.302151 242.404932
-113.262237 -24.824752 239.874068

the outputs
0.499338 0.499058 0.395562
0.491947 0.477063 0.810679
0.336357 0.462832 0.808338

the activations:
-220.263874 -462.167483 -754.781147
-107.457481 -238.177673 -164.617658

-238.384906 -343.044127 -313.799639

the outputs
0.210555 0.0588 0.01068
0.344175 0.193244 0.271365
0.19305 0.11322 0.132068

tourcity 0 tour order 0

tourcity 1 tour order 2

tourcity 2 tour order 1

the tour :
0
2
1

distance of tour is : 31

Output for a Four-City Problem

Please type number of cities, number of iterations
4   30

0.001027 0.000385 0.033491 0.003294   //input vector—there are 16
                                                neurons in the network.
0.035586 0.021728 0.053679 0.019547
0.07006 0.094996 0.02746 0.04444
0.010864 0.069846 0.056416 0.004149

type distance from city 0 to city 1
10

type distance from city 0 to city 2
14

type distance from city 0 to city 3
7

type distance from city 1 to city 2
6

type distance from city 1 to city 3
12

type distance from city 2 to city 3
9


```
0  10 14 7
10  0  6 12
14  6  0  9
7  12 9  0
```

```
-30  -70  -70  -70
-70  -30  -30  -30
-70  -30  -30  -30
-70  -30  -30  -30
```

```
-670  -30  -670  -70
-30   -70  -30   -30
-30   -70  -30   -30
-30   -70  -30   -30
```

```
-70  -910  -30  -910
-30  -30   -70  -30
-30  -30   -70  -30
-30  -30   -70  -30
```

```
-490  -70  -490  -30
-30   -30  -30   -70
-30   -30  -30   -70
-30   -30  -30   -70
```

```
-70  -30   -30  -30
-30  -670  -70  -670
-70  -30   -30  -30
-70  -30   -30  -30
```

```
-30  -70  -30  -30
-70  -30  -70  -70
-30  -70  -30  -30
-30  -70  -30  -30
```

```
-30  -30   -70  -30
-70  -430  -30  -430
-30  -30   -70  -30
-30  -30   -70  -30
```

```
-30   -30  -30   -70
-790  -70  -790  -30
-30   -30  -30   -70
-30   -30  -30   -70
```

```
-70 -30 -30 -30
-70 -30 -30 -30
-30 -910 -70 -910
-70 -30 -30 -30

-30 -70 -30 -30
-30 -70 -30 -30
-430 -30 -430 -70
-30 -70 -30 -30

-30 -30 -70 -30
-30 -30 -70 -30
-70 -70 -30 -70
-30 -30 -70 -30

-30 -30 -30 -70
-30 -30 -30 -70
-610 -70 -610 -30
-30 -30 -30 -70

-70 -30 -30 -30
-70 -30 -30 -30
-70 -30 -30 -30
-30 -490 -70 -490

-30 -70 -30 -30
-30 -70 -30 -30
-30 -70 -30 -30
-790 -30 -790 -70

-30 -30 -70 -30
-30 -30 -70 -30
-30 -30 -70 -30
-70 -610 -30 -610

-30 -30 -30 -70
-30 -30 -30 -70
-30 -30 -30 -70
-70 -70 -70 -30

the activations:
-0.739401 -0.739914 -80.377417 -5.138392
-68.325885 -15.644448 -77.297451 -42.221625
-168.144739 -136.794457 -19.771052 -79.992216
-16.948486 -150.868332 -101.549102 -2.987476

the outputs
0.498891 0.49889 0.381718 0.492293
0.398923 0.476551 0.386088 0.437004
0.267201 0.305601 0.470378 0.382263
0.474599 0.287981 0.35222 0.495519
```

the activations:
-105.306867 -223.608119 -401.389268 -461.851763
-227.545843 -383.173889 -624.165242 -791.883759
-253.311059 -378.545908 -439.022651 -632.30711
-163.485117 -438.905692 -574.849578 -673.718697

the outputs
0.347093 0.207239 0.082539 0.058905
0.203384 0.091202 0.02309 0.008566
0.179481 0.09353 0.066974 0.022013
0.272711 0.067017 0.030796 0.017254

tourcity 0 tour order 0

tourcity 1 tour order 1

tourcity 2 tour order 2

tourcity 3 tour order 3

the tour :
0
1
2
3

distance of tour is : 32

Other Efforts for Solving the Traveling Salesperson Problem

Anzai's Presentation

Anzai describes the Hopfield network for the traveling salesperson problem in a slightly different way. For one thing, a global inhibition term is not used. A threshold value is associated with each neuron, added to the activation, and taken as the average of A_1 and A_2, using our earlier notation. The **Energy** function is formulated slightly differently, as follows.

$$E = A_1 \sum_i (\sum_k x_{ik} - 1)^2 + A_2 \sum_i (\sum_k x_{ki} - 1)^2 +$$
$$A_4 \sum_k \sum_{j \neq k} \sum_{i \neq k} d_{kj} x_{ki}(x_{j,i+1} + x_{j,i-1})$$

The first term is 0 if the sum of the outputs is 1 in each column. The same is true for the second term with regard to rows.

The output is calculated using a parameter λ, here called the *reference activation level*, as:

$$x_{ij} = (1 + \tan \tanh(a_{ij}/\lambda)) / 2$$

The parameters used are $A_1 = 1/2$, $A_2 = 1/2$, $A_4 = 1/2$, $\Delta t = 1$, $\tau = 1$, and $\lambda = 1$. An attempt is made to solve the problem for a tour of 10 cities. The solution obtained is not crisp, in the sense that exactly one 1 occurs in each row and each column, there are values of varying magnitude with one dominating value in each row and column. The prominent value is considered to be part of the solution.

Kohonen's Approach for the Traveling Salesperson Problem

Kohonen's self-organizing maps can be used for the traveling salesperson problem. We summarize the discussion of this approach described in Eric Davalo and Patrick Naim's work. Each city considered for the tour is referenced by its x and y coordinates. To each city there corresponds a neuron. The neurons are placed in a single array, unlike the two-dimensional array used in the Hopfield approach. The first and the last neurons in the array are considered to be neighbors.

There is a weight vector for each neuron, and it also has two components. The weight vector of a neuron is the image of the neuron in the map, which is sought to self-organize. There are as many input vectors as there are cities, and the coordinate pair of a city constitutes an input vector. A neuron with a weight vector closest to the input vector is selected. The weights of neurons in a neighborhood of the selected neuron are modified, others are not. A gradually reducing scale factor is also used for the modification of weights.

One neuron is created first, and its weight vector has 0 for its components. Other neurons are created one at a time, at each iteration of learning. Neurons may also be destroyed. The creation of the neuron and destruction of the neuron suggest adding a city provisionally to the final list in the tour and dropping a city also provisionally from that list. Thus the possibility of assigning any neuron to two inputs or two cities is prevented. The same is true about assigning two neurons to the same input.

As the input vectors are presented to the network, if an unselected neuron falls in the neighborhood of the closest twice, it is created. If a created neuron is not selected in three consecutive iterations for modification of weights, along with others being modified, it is destroyed.

That a tour of shortest distance results from this network operation is apparent from the fact that the closest neurons are selected. It is reported that experimental results are very promising. The computation time is small and solutions somewhat close to the optimal are obtained, if not the optimal solution itself. As has been said before about the traveling salesperson problem, this is a NP-complete problem and near efficient approaches to it should be accepted.

Algorithm for Kohonen's Approach

A gain parameter λ and a scale factor q are used while modifying the weights. A value between 0.02 and 0.2 was tried for q. A distance of a neuron from the selected neuron is defined to be an integer between 0 and n − 1, where n is the number of cities for the tour. This means that these distances are not necessarily the actual distances between the cities. They could be made representative of the actual distances in some way. One such attempt is described in the discussion on C++ implementation that follows. This distance is denoted by d_j for neuron j. A **Squashing** function similar to the **Gaussian Density** function is also used.

The details of the algorithm are in a paper referenced in Davalo. The steps of the algorithm to the extent given by Davalo are:

◆ Find the weight vector for which the distance from the input vector is the smallest

◆ Modify the weights using

$$w_{jnew} = w_{jold} + (I_{new} - w_{jold})g(\lambda, d_j),$$
$$\text{where } g(\lambda, dj) = \exp(- dj^2/\lambda) /\sqrt{2}$$

♦ Reset λ as $\lambda(1 - q)$

C++ Implementation of Kohonen's Approach

Our C++ implementation of this algorithm (described above) is with small modifications. We create but do not destroy neurons explicitly. That is, we do not count the number of consecutive iterations in which a neuron is not selected for modification of weights. This is a consequence of our not defining a neighborhood of a neuron. Our example is for a problem with five neurons, for illustration, and because of the small number of neurons involved, the entire set is considered a neighborhood of each neuron.

When all but one neuron are created, the remaining neuron is created without any more work with the algorithm, and it is assigned to the input, which isn't corresponded yet to a neuron. After creating n – 1 neurons, only one unassigned input should remain.

In our C++ implementation, the distance matrix for the distances between neurons, in our example, is given as follows, following the stipulation in the algorithm that these values should be integers between 0 and n – 1.

```
        0 1 2 3 4
        1 0 1 2 3
d  =    2 1 0 1 2
        3 2 1 0 1
        4 3 2 1 0
```

We also ran the program by replacing the matrix above with the following matrix and obtained the same solution. The actual distances between the cities are about four times the corresponding values in this matrix, more or less. We have not included the output from this second run of the program.

```
        0 1 3 3 2
        1 0 3 2 1
d  =    3 3 0 4 2
        3 2 4 0 1
        2 1 2 1 0
```

In our implementation we picked a function similar to the **Gaussian Density** function as: the **Squashing** function. The **Squashing** function used is:

$$f(d,\lambda) = \exp(-d^2/\lambda) \; / \; \sqrt{(2\pi)}$$

Header File for C++ Program for Kohonen's Approach

Listing 15.3 contains the header file for this program. Listing 15.4 contains the corresponding source file.

Listing 15.3 Header file for C++ program for Kohonen's approach

```
//tsp_kohn.h V.Rao, H.Rao
#include<iostream.h>
#include<math.h>

#define MXSIZ 10
#define pi 3.141592654

class city_neuron
    {
     protected:
        double x,y;
        int mark,order,count;
        double weight[2];
        friend class tspnetwork;
     public:
        city_neuron(){};
        void get_neuron(double,double);
    };

class tspnetwork
    {
    protected:
        int chosen_city,order[MXSIZ];
        double gain,input[MXSIZ][2];
        int citycount,index,d[MXSIZ][MXSIZ];
        double gain_factor,diffsq[MXSIZ];
        city_neuron (cnrn)[MXSIZ];
```

```
public:
    tspnetwork(int,double,double,double,double*,double*);
    void get_input(double*,double*);
    void get_d();
    void find_tour();
    void associate_city();
    void modify_weights(int,int);
    double wtchange(int,int,double,double);
    void print_d();
    void print_input();
    void print_weights();
    void print_tour();
};
```

Source File Listing

The following is the source file listing for the Kohnen approach for the Traveling Salesperson Problem.

Listing 15.4 Source file for C++ program for Kohonen,s approach

```
//tsp_kohn.cpp  V.Rao, H.Rao
#include "tsp_kohn.h"

void city_neuron::get_neuron(double a,double b)
    {
    x = a;
    y = b;
    mark = 0;
    count = 0;
    weight[0] = 0.0;
    weight[1] = 0.0;
    };

tspnetwork::tspnetwork(int k,double f,double q,double h,dou-
    ble *ip0,double *ip1)
    {
    int i;
```

```
        gain = h;
        gain_factor = f;
        citycount = k;

        // distances between neurons as integers between 0 and n-1

        get_d();
        print_d();
        cout<<"\n";

        // input vectors

        get_input(ip0,ip1);
        print_input();

        // neurons in the network

        for(i=0;i<citycount;++i)
            {
            order[i] = citycount+1;
            diffsq[i] = q;
            cnrn[i].get_neuron(ip0[i],ip1[i]);
            cnrn[i].order = citycount +1;
            }
        }

void tspnetwork::associate_city()
    {
    int i,k,j,u;
    double r,s;

    for(u=0;u<citycount;u++)
        {
        //start a new iteration with the input vectors
        for(j=0;j<citycount;j++)
            {
            for(i=0;i<citycount;++i)
                {
```

```
            if(cnrn[i].mark==0)
                {
                k = i;
                i =citycount;
                }
            }

        //find the closest neuron

        for(i=0;i<citycount;++i)
            {
            r = input[j][0] - cnrn[i].weight[0];
            s = input[j][1] - cnrn[i].weight[1];
            diffsq[i] = r*r +s*s;
            if(diffsq[i]<diffsq[k]) k=i;
            }

        chosen_city = k;
        cnrn[k].count++;
        if((cnrn[k].mark<1)&&(cnrn[k].count==2))
            {
            //associate a neuron with a position
            cnrn[k].mark = 1;
            cnrn[k].order = u;
            order[u] = chosen_city;
            index = j;
            gain *= gain_factor;

            //modify weights
            modify_weights(k,index);
            print_weights();
            j = citycount;
            }
        }
    }
}
```

```
void tspnetwork::find_tour()
   {
   int i;
   for(i=0;i<citycount;++i)
      {
      associate_city();
      }

      //associate the last neuron with remaining position in
      // tour
      for(i=0;i<citycount;++i)
         {
         if( cnrn[i].mark ==0)
            {
            cnrn[i].order = citycount-1;
            order[citycount-1] = i;
            cnrn[i].mark = 1;
            }
         }

      //print out the tour.
      //First the neurons in the tour order
      // Next cities in the tour
      //order with their x,y coordinates

      print_tour();
      }

void tspnetwork::get_input(double *p,double *q)
   {
   int i;

   for(i=0;i<citycount;++i)
      {
      input[i][0] = p[i];
      input[i][1] = q[i];
      }
   }
```

```
//function to compute distances (between 0 and n-1) between
//neurons

void tspnetwork::get_d()
    {
    int i,j;

    for(i=0;i<citycount;++i)
        {
        for(j=0;j<citycount;++j)
            {
            d[i][j] = (j-i);
            if(d[i][j]<0) d[i][j] = d[j][i];
            }
        }
    }
```

```
//function to find the change in weight component

double tspnetwork::wtchange(int m,int l,double g,double h)
    {
    double r;
    r = exp(-d[m][l]*d[m][l]/gain);
    r *= (g-h)/sqrt(2*pi);
    return r;
    }
```

```
//function to determine new weights

void tspnetwork::modify_weights(int jj,int j)
    {
    int i;
    double t;
    double w[2];
    for(i=0;i<citycount;++i)
        {
        w[0] = cnrn[i].weight[0];
```

```
    w[1] = cnrn[i].weight[1];
    //determine new first component of weight
    t = wtchange(jj,i,input[j][0],w[0]);
    w[0] = cnrn[i].weight[0] +t;
    cnrn[i].weight[0] = w[0];

    //determine new second component of weight
    t = wtchange(jj,i,input[j][1],w[1]);
    w[1] = cnrn[i].weight[1] +t;
    cnrn[i].weight[1] = w[1];
    }
  }

//different print routines

void tspnetwork::print_d()
  {
  int i,j;
  cout<<"\n";

  for(i=0;i<citycount;i++)
    {
    cout<<" d: ";
    for(j=0;j<citycount;j++)
      {

      cout<<d[i][j]<<"    ";
      }
    cout<<"\n";
    }
  }

void tspnetwork::print_input()
  {
  int i,j;

  for(i=0;i<citycount;i++)
    {
```

```
        cout<<"input : ";
        for(j=0;j<2;j++)
            {
            cout<<input [i][j]<<"    ";
            }
        cout<<"\n";
        }
    }

void tspnetwork::print_weights()
    {
    int i,j;
    cout<<"\n";

    for(i=0;i<citycount;i++)
        {
        cout<<" weight: ";
        for(j=0;j<2;j++)
            {
            cout<<cnrn[i].weight[j]<<"    ";
            }
        cout<<"\n";
        }
    }

void tspnetwork::print_tour()
    {
    int i,j;
    cout<<"\n tour : ";

    for(i=0;i<citycount;++i)
        {
        cout<<order[i]<<" --> ";
        }
    cout<<order[0]<<"\n\n";

    for(i=0;i<citycount;++i)
        {
```

```
        j = order[i];
        cout<<"("<<cnrn[j].x<<", "<<cnrn[j].y<<") --> ";
        }
    j= order[0];
    cout<<"("<<cnrn[j].x<<", "<<cnrn[j].y<<")\n\n";
    }

void main()
    {
    int nc= 5;//nc = number of cities
    double q= 0.05,h= 1.0,p= 1000.0;

    double input2[][5]=
    {7.0,4.0,14.0,0.0,5.0,3.0,6.0,13.0,12.0,10.0};
    tspnetwork tspn2(nc,q,p,h,input2[0],input2[1]);
    tspn2.find_tour();


    }
```

Output from a Sample Program Run

The program, as mentioned, is created for the Kohonen approach to the traveling salesperson problem for five cities. There is no user input from the keyboard. All parameter values are given to the program with appropriate statements in the function main. A scale factor of 0.05 is given to apply to the gain parameter, which is given as 1. Initially, the distance of each neuron weight vector from an input vector is set at 1000, to facilitate finding the closest for the first time. The cities with coordinates (7,3), (4,6), (14,13), (0,12), (5,10) are specified for input vectors.

The output of the program being all computer generated is given in the following listing in boldface. The tour found is not the one in natural order, namely $0 \rightarrow 1 \rightarrow 2 \rightarrow 3 \rightarrow 4 \rightarrow 0$, with a distance of 43.16. The tour found has the order $0 \rightarrow 3 \rightarrow 1 \rightarrow 4 \rightarrow 2 \rightarrow 0$, which covers a distance of 44.43, which is slightly higher. The best tour, $0 \rightarrow 2 \rightarrow 4 \rightarrow 3 \rightarrow 1 \rightarrow 0$ has a total distance of 38.54.

Table 15.2 gives for the five-city example, the twelve (5/10) distinct tour distances and corresponding representative tours. These are not generated by the program, but by enumeration and calculation by hand. This table is provided here for you to see the different solutions for this five-city example of the traveling salesperson problem.

TABLE 15.2 *Distances and representative tours for five-city example.*

Distance	Tour	Comment
49.05	0 - 3 - 2 - 1 - 4 - 0	worst case
47.59	0 - 3 - 1 - 2 - 4 - 0	
45.33	0 - 2 - 1 - 4 - 3 - 0	
44.86	0 - 2 - 3 - 1 - 4 - 0	
44.43	0 - 3 - 1 - 4 - 2 - 0	tour given by the program
44.30	0 - 2 - 1 - 3 - 4 - 0	
43.29	0 - 1 - 4 - 2 - 3 - 0	
43.16	0 - 1 - 2 - 3 - 4 - 0	tour in natural order
42.73	0 - 1 - 2 - 4 - 3 - 0	
42.26	0 - 1 - 3 - 2 - 4 - 0	
40.00	0 - 1 - 4 - 3 - 2 - 0	
38.54	0 - 2 - 4 - 3 - 1 - 0	optimal tour

There are twelve different distances you can get for tours with these cities by hand calculation, and four of these are higher and seven are lower than the one you find from this program. The worst case tour $(0 \rightarrow 3 \rightarrow 2 \rightarrow 1 \rightarrow 4 \rightarrow 0)$ gives a distance of 49.05, and the best, as you saw above, 38.54. The solution from the program is at about the middle of the best and worst, in terms of total distance traveled. It is a good result, even if it is not the best.

```
d: 0  1  2  3  4
d: 1  0  1  2  3
d: 2  1  0  1  2
d: 3  2  1  0  1
d: 4  3  2  1  0

input : 7  3
input : 4  6
```

```
input : 14  13
input : 0  12
input : 5  10

weight: 1.595769  2.393654
weight: 3.289125e-09  4.933688e-09
weight: 2.880126e-35  4.320189e-35
weight: 1.071429e-78  1.607143e-78
weight: 1.693308e-139  2.539961e-139

weight: 1.595769  2.393654
weight: 5.585192  5.18625
weight: 2.880126e-35  4.320189e-35
weight: 1.071429e-78  1.607143e-78
weight: 1.693308e-139  2.539961e-139

weight: 1.595769  2.393654
weight: 5.585192  5.18625
weight: 5.585192  5.18625
weight: 1.071429e-78  1.607143e-78
weight: 1.693308e-139  2.539961e-139

weight: 1.595769  2.393654
weight: 5.585192  5.18625
weight: 5.585192  5.18625
weight: 5.585192  5.18625
weight: 1.693308e-139  2.539961e-139

weight: 1.595769  2.393654
weight: 5.585192  5.18625
weight: 5.585192  5.18625
weight: 5.585192  5.18625
weight: 5.585192  5.18625

tour : 0 → 3→ 1 → 4 → 2→ 0

(7, 3) → (0, 12) → (4, 6) → (5, 10) → (14, 13) → (7, 3)
```

Summary

The traveling salesperson problem is presented in this chapter as an example of nonlinear optimization with neural networks. Details of formulation of the **energy** function and evaluation of it are given. The approaches to the solution of the traveling salesperson problem using a Hopfield network and using a Kohonen self-organizing map are presented. C++ programs for both approaches are included.

The output with the C++ program for the Hopfield network refers to examples of three- and four-city tours. The output with the C++ program for the Kohonen approach is given for a tour of five cities, for illustration. The solution obtained is a good one, if not the optimal. The problem with the Hopfield approach lies in the selection of appropriate values for the parameters. Hopfield's choices for his 10-city tour problem are given. The same values for the parameters may not work for the case of a different number of cities. The version of this approach given by Anzai is also discussed briefly.

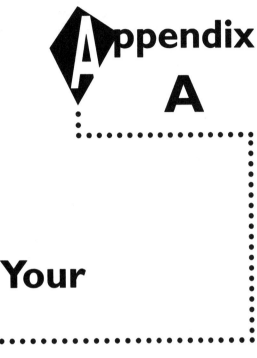

Appendix
A

Compiling Your Programs

ll of the programs included in the book have been compiled and tested on Turbo C++, Borland C++, and Microsoft C++/Visual C++ with either the small or medium memory model. You should not have any problems in using other compilers, since standard I/O routines are used. With the backpropagation simulator of Chapters 7, 13, and 14 you may run into a memory shortage situation. You should unload any TSR (Terminate and Stay Resident) programs and/or choose smaller architectures for your networks. By going to more hidden layers with fewer neurons per layer, you may be able to reduce the overall memory requirements.

The programs in this book make heavy use of floating point calculations, and you should compile your programs to take advantage of a math coprocessor, if you have one installed in your computer.

The organization of files on the accompanying diskette are according to chapter number. You will find relevant versions of files in the corresponding chapter directory.

379

Most of the files are self-contained, or include other needed files in them. Load the main file for example for backpropagation, the backprop.cpp file, into the development environment editor for your compiler and build a .exe file. That's it!

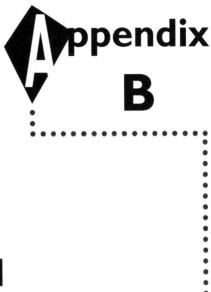

Appendix B

Mathematical Background

Dot Product or Scalar Product of Two Vectors

iven vectors U and V, where $U = (u_1, ..., u_n)$ and $V = (v_1, ..., v_n)$, their dot product or scalar product is $U \cdot V = u_1v_1 + ... + u_nv_n$.

Matrices and Some Arithmetic Operations on Matrices

A real matrix is a rectangular array of real numbers. A matrix with m rows and n columns is referred to as an m x n matrix. The element in the ith row and jth column of the matrix is referred to as the ij element of the matrix and is typically denoted by a_{ij}.

The transpose of a matrix M is denoted by M^T. The element in the ith row and jth column of M^T is the same as the element of M in its jth row and ith column. M^T is obtained from M by interchanging the rows and columns of M. For example, if

$$M = \begin{matrix} 2 & 7 & -3 \\ 4 & 0 & 9 \end{matrix} \text{, then } M^T = \begin{matrix} 2 & 4 \\ 7 & 0 \\ -3 & 9 \end{matrix}$$

If X is a vector with m components, $x_1, ..., x_m$, then it can be written as a column vector with components listed one below another. It can be written as a row vector, $(x_1, ..., x_m)$. The transpose of a row vector is the column vector with the same components, and the transpose of a column vector is the corresponding row vector.

The addition of matrices is possible if they have the same size, that is, the same number of rows and same number of columns. Then you just add the ij elements of the two matrices to get the ij element of the sum matrix. For example,

$$\begin{matrix} 3 & -4 & 5 \\ 2 & 3 & 7 \end{matrix} + \begin{matrix} 5 & 2 & -3 \\ 6 & 0 & 4 \end{matrix} = \begin{matrix} 8 & -2 & 2 \\ 8 & 3 & 11 \end{matrix}$$

Multiplication is defined for a given pair of matrices, only if a condition on their respective sizes is satisfied. Then too, it is not a commutative operation. This means that if you exchange the matrix on the left with the matrix on the right, the multiplication operation may not be defined, and even if it is, the result may not be the same as before such an exchange.

The condition to be satisfied for multiplying the matrices A, B as AB is, that the number of columns in A is equal to the number of rows in B. Then to get the ij element of the product matrix AB, you take the ith row of A as one vector and the jth column of B as a second vector and do a dot product of the two. For example, the two matrices given previously to illustrate the addition of two matrices are not compatible for multiplication in whichever order you take them. It is because there are three columns in each, which is different from the number of rows, which is 2 in each. Another example is given as follows.

$$\text{Let } A = \begin{array}{ccc} 3 & -4 & 5 \\ 2 & 3 & 7 \end{array} \quad \text{and } B = \begin{array}{cc} 5 & 6 \\ 2 & 0 \\ -3 & 4 \end{array}$$

Then AB and BA are both defined, AB is a 2 x 2 matrix, whereas BA is 3 x 3.

$$\text{Also } AB = \begin{array}{cc} -8 & 38 \\ -5 & 40 \end{array} \quad \text{and } BA = \begin{array}{ccc} 27 & -2 & 67 \\ 6 & -8 & 10 \\ -1 & 24 & 13 \end{array}$$

Lyapunov Function

A Lyapunov function is a function that does not take negative values and has the property of making a correspondence between the state variables of a system and real numbers, and decreasing with time. The state of the system changes as time changes, and the function decreases. Thus, the Lyapunov function decreases with each change of state of the system.

We can construct a simple example of a function with the property of decreasing with each change of state as follows.

Suppose a real number x, represents the state of a dynamic system at time t. Also suppose that x is bounded for any t by a positive real number M. That means x is less than M for every value of t.

Then the function,

```
f(x,t) = exp(-|x|/(M+|x|+t))
```

is nonnegative and decreases with increasing t.

Local Minimum

A function f(x) is defined to have a local minimum at y, with a value z, if

f(y) = z, and f(x) ≥ z, for each x, such that there exists a positive real number h such that y – h ≤ x ≤ y + h.

In other words, there is no other value of x in a neighborhood of y, where the value of the function is smaller than z.

There can be more than one local minimum for a function in its domain. A **Step** function (with a graph resembling a stair case) is a simple example of a function with an infinite number of points in its domain with local minima.

Global Minimum

A function f(x) is defined to have a global minimum at y, with a value z, if

f(y) = z, and f(x) ≥ z, for each x in the domain of the function f.

In other words, there is no other value of x in the domain of the function f, where the value of the function is smaller than z. Clearly, a global minimum is also a local minimum, but a local minimum may not be a global minimum.

There can be more than one global minimum for a function in its domain. The trigonometric function f(x) = sinx is a simple example of a function with an infinite number of points with global minima. You may recall that sin(3π/ 2), sin (7π/ 2), and so on are all –1, the smallest value for the sine function.

Kronecker Delta Function

The Kronecker delta function is a function of two variables. It has a value of 1 if the two arguments are equal, and 0 if they are not. Formally,

$$\delta(x,y) = \begin{cases} 1 \text{ if } x = y \\ 0 \text{ if } x \neq y \end{cases}$$

Gaussian Density Distribution

The Gaussian Density distribution, also called the Normal distribution, has a density function of the following form. There is a constant parameter c, which can have any positive value.

$$f(x) = \frac{1}{\sqrt{(2\pi c)}} \exp(-x^2 / c)$$

Glossary

A

Activation	The weighted sum of the inputs to a neuron in a neural network.
Adaline	Adaptive Linear Element Machine.
Adaptive Resonance Theory	Theory developed by Grossberg and Carpenter for categorization of patterns, and to address the stability–plasticity dilemma.
Algorithm	A step-by-step procedure to solve a problem.
Annealing	A process for preventing a network from being drawn into a local minimum.
ART	(Adaptive Resonance Theory) ART1 is the result of the initial development of this theory for binary inputs. Further developments led to ART2 for analog inputs. ART3 is the latest.
Artificial neuron	The primary object in a neural network of human creation, to mimic the neuron activity of the brain. It is a processing element of a network.
Associative memory	Activity of associating one pattern or object with itself or another.

387

Autoassociative   Making a correspondence of one pattern or object with itself.

B

Backpropagation   A neural network model where the errors at the output layer are propagated back to the layer before in learning. If the previous layer is not the input layer, then the errors at this hidden layer are propagated back to the layer before.

BAM   Bidirectional Associative Memory network model.

Bias   A value added to the activation of a neuron.

Binary digit   A value of 0 or 1.

Bipolar value   A value of –1 or +1.

Boltzman Machine   A neural network in which the outputs are determined with probability distributions. Well suited for simulated annealing process.

Brain-State-in-a-Box   Anderson's single-layer, laterally connected neural network model. It can work with inputs that have noise in them or which are incomplete.

C

Cauchy Machine   Similar to the Boltzman Machine, except that a Cauchy distribution is used for probabilities.

Cognitron   The forerunner in the development of Neocognitron. A network developed to recognize characters.

Competition   A process in which a winner is selected from a layer of neurons by some criterion. Competition suggests inhibition reflected in some connection weights being assigned a negative value.

Connection	A means of passing inputs from one neuron to another.
Connection weight	A numerical label associated with a connection and used in a weighted sum of inputs.
Constraint	A condition expressed as an equation or inequality, which has to be satisfied by the variables.
Convergence	Termination of a process naturally with a final result.

D

Delta rule	A rule for modification of connection weights, using both the output and the error obtained. It is also called the *LMS rule*.

E

Energy function	A function of outputs and weights in a neural network to determine the state of the system.
Excitation	Providing positive weights on connections to enable outputs that cause a neuron to fire.
Exemplar	An example of a pattern or object used in training a neural network.
Expert System	A set of formalized rules that help to perform like an expert.

F

FAM	Fuzzy Associative Memory network. Makes associations between fuzzy sets.
Feedback	The process of relaying information in the opposite direction to the original.

Fit vector A vector of values of degree of membership of elements of a fuzzy set.

Fully Connected Network A neural network in which every neuron has connections to all other neurons.

Fuzziness Different concepts having an overlap to some extent. For example, descriptions of fair and cool temperatures may have an overlap of a small interval of temperatures.

Fuzzy Associative Memory A neural network model to make association between fuzzy sets.

G

Gain Sometimes a numerical factor to enhance the activation. Sometimes a connection for the same purpose.

Generalized Delta rule A rule used in networks such as Backpropagation where hidden layer weights are modified with backpropagated error.

Global Minimum A point where the value of a function is no greater than the value at any other point in the domain of the function.

H

Hamming Distance The number of places in which two bit vectors differ from each other.

Hebbian Learning A learning algorithm in which Hebb's rule is used. The change in connection weight between two neurons is taken as a constant times the product of their outputs.

Heteroassociative Making an association between two distinct patterns or objects.

Hidden Layer An array of neurons positioned in between the input and output layers.

Hopfield Network   A single layer fully connected autoassociative neural network.

I

Inhibition   The attempt by one neuron to diminish the chances of firing by another neuron.

Input layer   An array of neurons to which an external input or signal is presented.

Instar   A neuron that has no connections going from it to other neurons.

Iteration   Repetition of a procedure. A cycle of processing in a neural network.

L

Lateral connection   A connection between two neurons that belong to the same layer.

Layer   An array of neurons positioned similarly in a network for its operation.

Learning   The process of finding an appropriate set of connection weights to achieve the goal of the network operation.

Linearly separable   Two subsets of a set having a linear barrier between the two of them.

LMS rule   Least mean squared error rule, with the aim of minimizing the average of the squared error. Same as the Delta rule.

Local minimum   A point where the value of the function is no greater than the value at any other point in its neighborhood.

Long-Term Memory (LTM)   Encoded information that is retained for an extended period.

Lyapounov function

A function that is bounded and represents the state of a system and that decreases with every change in the state of the system.

M

Madaline

A neural network in which the input layer has units that are Adalines. It is a multiple-adaline.

Mapping

A correspondence between elements of two sets.

N

Neural Network

A collection of processing elements arranged in layers, and a collection of connection edges between pairs of neurons. Input is received at one layer, and output is produced at the same or at a different layer.

Noise

Distortion of an input.

Nonlinear Optimization

Finding the best solution for a problem that has a **nonlinear** function in its objective or in a constraint.

O

On Center Off Surround

Assignment of excitatory weights to connections to nearby neurons and inhibitory weights to connections to distant neurons.

Orthogonal vectors

Vectors whose dot product is 0.

Outstar

A neuron that has no incoming connections.

P

Perceptron

A neural network for linear pattern matching.

Plasticity	Ability to be stimulated by new inputs.

R

Resonance	The responsiveness of two neurons in different layers in categorizing an input. An equilibrium in two directions.

S

Saturation	A condition of limitation on the frequency with which a neuron can fire.
Self-Organization	A process of partitioning the output layer neurons to correspond to individual patterns or categories.
Short-Term Memory (STM)	The storage of information that does not endure long after removal of the corresponding input.
Simulated annealing	An algorithm by which changes are made to decrease energy or temperature or cost.
Stability	Convergence of a network operation to a stable solution.
Supervised learning	A learning process where exemplars are used.

T

Threshold value	A value used to compare the output of a neuron to determine if the neuron fires or not. Sometimes a threshold value is added to the activation of a neuron to determine its output.
Training	The process of helping in learning by a neural network by providing desired values corresponding to inputs.

U

Unsupervised Learning Learning in the absence of external information on outputs.

V

Vigilance parameter A parameter used in Adaptive Resonance Theory. It is used to determine if the activation of a subsystem in the network is to be prevented.

W

Weight A number associated with a neuron or with a connection between two neurons, which is used in aggregating the outputs to determine the activation of a neuron.

References

Aleksander, Igor, and Morton, Helen, *An Introduction to Neural Computing*, Chapman and Hall, London, 1990.

Anzai, Yuichiro, *Pattern Recognition and Machine Learning*, Academic Press, Englewood Cliffs, N.J. 1992.

Davalo, Eric, and Naim, Patrick, *Neural Networks*, MacMillan, New York, 1991.

Freeman, James A. and Skapura, David M., *Neural Networks Algorithms, Applications, and Programming Techniques*, Addison-Wesley, Reading, MA., 1991.

Grossberg, Stephen et al., *Introduction and Foundations*, Lecture Notes, Neural Network Courses and Conference, Boston University, May 1992.

Hammerstrom, Dan, *Neural Networks at Work*, IEEE Spectrum, New York, June 1993.

Hertz, John, Krogh, Anders, and Palmer, Richard, *Introduction to the theory of neural computation*, Addison-Wesley, Reading, MA., 1991.

Johnson, R. Colin, *Making the Neural-Fuzzy Connection*, Electonic Engineering Times, CMP Publications, Manhasset, N.Y., September 27, 1993.

Jurik, Mark, "The care and feeding of a neural network," *Futures Magazine,* Oster Communications, Cedar Falls, Iowa, October 1992.

Kline, J. and Folger, T.A., *Fuzzy Sets, Uncertainty and Information,* Prentice Hall, New York, 1988.

Kosko, Bart and Isaka, Satoru, *Fuzzy Logic,* Scientific American, New York, July 1993.

Kosko, Bart, *Neural Networks And Fuzzy Systems: A Dynamical Systems Approach to Machine Intelligence,* Prentice-Hall , New York, 1992.

MacGregor, Ronald J., *Neural and Brain Modeling,* Academic Press, Englewood Cliffs, N.J., 1987.

Maren, Alianna, Harston, Craig, and Pap, Robert, *Handbook of Neural Computing Applications,* Academic Press, Englewood Cliffs, N.J., 1990.

Murphy, John J., *Intermarket Technical Analysis,* John Wiley and Sons, New York, 1991.

Rao, Valluru and Rao, Hayagriva, *Power Programming Turbo C++,* MIS:Press, New York, 1992.

Simpson, Patrick K., *Artificial Neural Systems: Foundations, Paradigms, Applications and Implementations,* Pergamon Press, London, 1990.

Soucek, Branko, and Soucek, Marina, *Neural and Massively Parallel Computers: The Sixth Generation,* John Wiley & Sons, New York, 1988.

Wasserman, Philip D., *Neural Computing,* Van Nostrand Reinhold, New York, 1989.

Yager, R, editor, *Fuzzy Sets and Applications: Selected Papers by L.Z. Zadeh,* Wiley-Interscience, New York, 1987.

Index

397

T

TSR, *see Terminate and Stay Resident*
Tank, 335
target
    outputs, 94, 96
    patterns, 88
tau, 346
temperature, 99, 100
Temporal Associative Memory, 76
Terminate and Stay Resident programs, 379
terminating value, 246
termination criterion, 265
test.dat file, 269
test mode, 111, 112, 143, 269
Thirty-year Treasury Bond Rate, 314, 317
Three-month Treasury Bill Rate, 314
threshold
    function, 2, 4, 44, 79, 83, 84, 105, 152-154, 204
    value, 10, 44, 61, 63, 74, 84, 106
thresholding, 72
    function, 110, 146
time lag, 316
time shifting, 324, 325
tolerance, 265, 268, 270, 272
    level, 63-64, 83, 103
    value, 99
top-down connections, 90, 200
top-down inputs, 203
Topology Preserving Maps, 97
total volume, 314
tour, 333
traces, 200
    of STM, 200
    of LTM, 200
trading volume, 314
training, 6, 40, 82, 90, 103, 228
    fast, 90
    law, 224, 250
    mode, 111, 114, 143

supervised, 61, 78
slow, 90
time, 271
unsupervised, 61
transpose
    of a matrix, 150, 152, 175, 197
    of a vector, 81
traveling salesperson(salesman) problem, 99, 333
    hand calculation, 336
    Hopfield network solution-Hopfield, 340
    Hopfield network solution-Anzai, 363
    Kohonen network solution, 364
triple, 179
truth value, 25
tsneuron class, 344
two-layer networks, 76
two-thirds rule, 90, 201, 204, 221

U

Unemployment Rate, 314
undertrained network, 272
union, 27
unit hypercube, 178
unit length, 225
universal set, 28
unsupervised, 90, 93
    learning, 6, 40, 93, 96, 97
    training, 61
uptrend, 314

V

value
    fit value, 27
    threshold value, 10, 44, 61, 63, 74, 84, 106
variable
    external, 23
    global, 23

How To Use Your Disk

The accompanying disk contains all of the source code for this book. Also, you have raw and processed data for the financial forecasting simulation described in Chapter 14. The files are organized according to the chapter in which they are used. There is a directory for each chapter that has source codes or data. Each chapter directory is self-contained and contains all the files you need to compile the programs for that chapter.

To compile any program, copy the files from the chapter directory to a directory of your hard disk. Then compile and build the primary .cpp file for that program in your working directory.